# The Conservator's Cookbook

*The Conservator's Cookbook* is a collection of practical, step-by-step guides outlining how to prepare various solutions, adhesives, gels, and other mixtures used in heritage conservation.

While most conservators learn the chemical rationale of solution-making during their training, the actual process of preparing these compounds can be neglected or eventually forgotten. This "cookbook" provides a non-exhaustive and adaptable resource, compiling practical recipes from across heritage conservation literature into a single volume. Drawing from techniques used in furniture, paper, paintings, textiles, and other conservation specialisms, each "recipe" in the Cookbook begins with a short summary of key chemical ideas, before relating step-by-step instructions. The solutions range from simple dissolutions (adhesive resins) to more complex cleaning systems (utilising surfactants, chelation agents, and enzymes) and gelled carriers. Each "recipe" is followed with collections of technical data which help to bridge the gap between chemical understanding and practical application. The *Cookbook* also provides important notes on health and safety and laboratory best practice, addressing some common misunderstandings and encouraging more sustainable approaches.

Intended for the experienced professional and conservation student alike, *The Conservator's Cookbook* is a crucial reference work that provides a starting point for practical experimentation and interaction between different conservation specialisms.

**Laura Chaillie** is a London-based objects conservator. Drawing from her background as an instructor, she is a keen communicator and active researcher. Her research interests include the characterisation and interventive care of plastics, with special focus on the preservation of electronic musical instruments.

# The Conservator's Cookbook

## Solution Preparation for the Heritage Professional

**Laura Chaillie**

Routledge
Taylor & Francis Group
LONDON AND NEW YORK

Designed cover image: coldsnowstorm / Getty Images

First published 2025
by Routledge
4 Park Square, Milton Park, Abingdon, Oxon OX14 4RN

and by Routledge
605 Third Avenue, New York, NY 10158

*Routledge is an imprint of the Taylor & Francis Group, an informa business*



*British Library Cataloguing-in-Publication Data*
A catalogue record for this book is available from the British Library

ISBN: 978-1-032-48979-7 (hbk)
ISBN: 978-1-032-48978-0 (pbk)
ISBN: 978-1-003-39170-8 (ebk)

DOI: 10.4324/9781003391708

Typeset in Times New Roman
by SPi Technologies India Pvt Ltd (Straive)

**The recipes provided in this book should only be undertaken by trained professionals.**

# Contents

# Chemical Structures

All structures were produced using the SMILES input function of the PubChem Sketcher produced by Ihlenfeldt, Bolton and Bryant (2009) and run through the PubChem website.

W. D. Ihlenfeldt, E. E. Bolton, and S. H. Bryant, "The PubChem chemical structure sketcher," *Journal of Cheminformatics*, vol. 1, no. 1, p. 20, Dec. 2009, doi: 10.1186/1758-2946-1-20.

PubChem, "PubChem Sketcher V2.4," PubChem Sketcher. Accessed: Jun. 08, 2024. [Online]. Available: https://pubchem.ncbi.nlm.nih.gov//edit3/index.html

# Figures

# Tables

# List of Consultants

**Steven Abbott, PhD**
Industrial Chemistry | Research Director | Professor
Steven Abbott TCNF Ltd.
0000-0002-7369-4408

**Sophie Barbisan**
Paper
Saint Louis Art Museum

**Megan Creamer**
Textiles
Art Institute of Chicago

**Jennifer Dinsmore**
Stone | Lecturer
City & Guilds of London Art School

**Joanne Hackett**
Textiles | Lecturer
University of Glasgow
0000-0002-5001-7797

**Thomas Higgins, PhD**
Inorganic Chemistry | Professor
Harold Washington College
0000-0003-4821-3171

**Carola Skov Hilby**
Paper
Munchmuseet

**Barry Knight, PhD**
Conservation Scientist
0000-0003-0555-0848

**Kyoko Kusunoki**
Japanese Paintings
British Museum
0000-0002-6874-8758

**Mathilde Renauld**
Paper and Photographs
Victoria and Albert Museum
0000-0003-4011-3977

**Shayne Rivers**
Furniture and Asian Lacquer | Head of School of Conservation | Lecturer
West Dean College

**Joyce H. Townsend, PhD**
Materials and Pigments, Paintings | Senior Conservation Scientist
Tate
0000-0003-4828-9264

**Astrid van Giffen**
Ceramics and Glass
Corning Museum of Glass

# Acknowledgements

To say that this wouldn't have been possible without the support of the listed consultants and myriad unlisted others would sprint straight past understatement and plunge headlong into absurdity. I've been bowled over by the openness of those that have so generously contributed their time and expertise to this little *Cookbook*.

It also would not have been possible without the unwavering support of the London Museum, in particular Catherine Nightingale and Sharon Robinson-Calver for their flexibility in scheduling and generosity with materials. I'd also like to thank my colleagues at the London Museum for providing a fantastic sounding board for ideas and techniques. I'll be sure to keep us flush with the good caramel biscuits for tea.

An extra emphatic thank-you needs to go to the two chemists I was lucky enough to work with. On my own, I know just enough chemistry to get really turned around: like trying to drive when you know a little bit about how a car works, but nothing about traffic, the highway code, or navigation. Much of the chemical nuance present in the first chapter is therefore the direct result of the corrections provided by Dr. Thomas Higgins. Similarly, the new methods of microemulsion formulation are only possible because Professor Abbott came to London and spent an afternoon shaking bottles with me. Thank you both.

Special thanks to Alex Gomez for graciously sharing some surfactant samples all the way from Texas, and Marcia Adams for her work with the project from day one.

Finally, the biggest thank-you goes to my brilliant partner, Mark. I can't thank you enough for your support, understanding, whinge deflection, endless cooking, and good humour during this whole process. You really are *the* best.

# Introduction

To be frank, I started compiling this text simply because it was a resource I wanted for my own emerging practice. In the years following my training, I'd repeatedly found that the search for practical information was time-consuming and, more often than not, concluded with a headlong sprint into a paywall.

The trouble was, I understood when and how to use something like an EDTA gel, but didn't know how to make it. What are useful concentrations? How hot does it need to be? When to add the chelation agent? What about buffering and pH adjustment? Would this destabilise the gel network?

I eventually decided to take full advantage of my alumni library access and start systematically compiling recipes, pulling practical details from an array of academic sources. I then figured, in what I've come to realise was a near-psychotic episode of optimism, that if I was going to be researching it, I might as well see if someone was interested in publishing it.

What I didn't immediately realise, but certainly know now, was just how far outside my own expertise this exercise would take me. The trouble with researching with an eye for personal enrichment, versus research for publication, is that the latter requires you not only to gather information, but to bring it back and present it. Even harder, you ideally present it in a way that is both comprehensible and accurate.

Given the scale and scope of this text, and the desire to finish it sometime within my working lifetime, I realised that I would need more than a little help. And so, I reached out to colleagues, authors, and experts from all over our wide discipline and beyond, requesting that they bring their sharpest editorial hatchet and hack this compilation into the form you now hold.

This project simply could not have happened without the guidance, clarification, and ever-growing bibliographic suggestions supplied by the consultants of this text. The words and the responsibility for accuracy are all mine, but the knowledge and depth of experience represented here reflect their expertise, and that of the cited authors, and not my own.

It is also worth making special mention of similar projects within the wider conservation discipline. There is a long tradition of recipes decks, with individual conservators or departments recording their preparations and tips. This tradition has also been carried over into digital formats with projects like the

DOI: 10.4324/9781003391708-1

conservation Wikis operated by the American Institute for Conservation (AIC). These crowdsourced Wikis can be excellent, but the level of engagement varies enormously, and the actual structure of the wiki can encourage the siloing of information within specialisms. For example, the fantastic Book and Paper Group Recipe wiki contains exactly the kind of practical advice I had been originally seeking out. And while this text has gone on to be a much larger project, it honestly likely wouldn't have started at all, had I known the BPG Wiki was there.

I have drawn directly from these and other Wikis and they are regularly cited throughout this text. I have also given specific credit to the source of the recipes being adapted and have endeavoured to flag up particularly important or relevant books, articles, and systems.

With all those musings and mea culpas behind us, a few short words on the text as a whole. Somewhat counterintuitively, it's worth first describing this text in terms of what it is not. *The Conservator's Cookbook* does not distil the complexity of conservation decision-making into one volume. In fact, only cursory attention is paid to chemical composition, and no attention at all is given to treatment rationale. When to employ which adhesive / cleaning solution / enzyme gel to what kind of object is vast, nuanced, and thoroughly covered in lengthy specialist literature.

This text, therefore, assumes a certain level of understanding about materials and the ethics of intervention. It also assumes a thorough knowledge of how new approaches must be carefully and systematically tested prior to use.

What this book does try to do is to compile some examples of practical solution preparation guidance from all over the conservation literature into a single resource. As previously stated, the aim isn't to provide any assistance with when to use which material, but simply to collect step-by-step instructions for how to make different solutions and preparations. This will ideally provide a starting point for conservators who understand methodology and materials but might lack the practical confidence to begin trialling new approaches.

The text is broken up into four larger sections, grouped by similarities in preparation and use rather than composition. As a result, many sections can work in tandem, such as gels and cleaning solutions, and where particularly relevant these connections will be indicated. Each recipe includes a short overview to provide chemical and contextual context and is followed by a collection of technical data. Occasionally, useful tips or techniques which are likely too pedestrian to make it to the professional literature are included, but as these have not been formally determined, they are therefore described as 'anecdotal'. The tone of this work is also quite casual and sometimes trades true scientific precision for general comprehensibility.

The 'Bench Skills' section includes some general guidelines on how to make the most of your laboratory equipment, some information about health and safety, and some useful calculations and tables.

As one might expect from the title, adhesives and consolidants includes preparation guidance for acrylic polymers, epoxies, dispersions, and natural glues and gums. This ranges from simple dilutions to the process of 'cooking' required by materials such as wheat starch paste. This is followed by a short chapter on rheology modifiers, which outlines a few natural and synthetic clays before launching into the materials and preparation of various spreadable and rigid gel systems.

'Cleaning', the largest section, doesn't have as many recipes as the previous one. Instead, it first outlines key chemical concepts, such as stability constants for chelation agents and then provides simple guidance for solution-making. Each 'recipe' is followed by tables of technical data, which are provided to help bridge the gap between complex theory and practical solutions.

While I have attempted to funnel as much information into this text as I could, it will be up to the individual conservator to turn that guidance into appropriate interventions for each specific object, condition, context, and use. This text sometimes includes non-archival materials (such as opaque epoxies), or defunct chemicals (such as some surfactants) as these can still be useful in certain context or as a reference to older literature.

*Inclusion or discussion of a material therefore does not automatically imply its suitability as a conservation material.* As ever, this must be assessed on a case-by-case, material-by-material basis.

Now, one final aside. It's worth being explicit that knowledge, or even mastery, of the solutions described in this text won't make you a conservator. I don't say this as some kind of hulking industry doorman, or in defence of academia, but simply to emphasise that the skill of the conservator isn't based in the solutions we make. Instead, it is rooted in the careful navigation of conflicting significances and the near-infinite complexity of treatment rationale. Knowledge of materials, as discussed in specialist texts, compliance with ethics, as dictated by professional bodies, the incorporation of expert knowledge, from cultural or authorial sources, are just a few of the many, many factors that inform treatment decision-making. These can't be found in one book, and certainly not this one book.

Instead, I hope this text is simply helpful. It is certainly not perfect and as it ages, the materials it lists and techniques it cites will no doubt be replaced with new formulations and approaches, but I hope that for now, it's made your professional life a little bit easier.

# 1 Bench Skills

## Overview

The working title for this section was originally 'Lab Skills' because many conservation activities, particularly the preparation of various solutions, are typically pictured within the controlled environment of a conservation laboratory. This image, however, reflects only one small slice of conservation activity and doesn't resonate with work done in field tents, museum basements, and back garden studios.

Before launching into the recipes, it seemed worthwhile to have a short chapter dedicated to the kinds of notes, guides, and reminders that enable accurate solution preparation. As such, this section will first address some key ideas and misconceptions about working within the capabilities of specific measuring equipment before highlighting a few useful calculations. The final section will focus on practical health and safety information with particular emphasis on eyewear, gloves, masks, solvent work, and regional resources.

Those interested in a more in-depth discussion of the scientific rationale behind conservation decisions are very strongly encouraged to seek out the "Science for Conservators" series. Of particular relevance is the first book *An Introduction to Materials and Chemistry*, newly revised by J. H. Townsend [1]. At the time of writing, it is very reasonably priced, particularly for the depth of information it carries, and would no doubt prove a worthwhile addition to any conservation space.

### *Online Retailers and Quality*

First, however, a quick aside. Anyone in this industry will be aware of the budgetary pressures and razor-thin margins that define the financial environment for the heritage sector. Conservators working in private practice and major institutions alike are under increased pressure to do more and bigger with less and less. As a result, there is an increased temptation to just opt for the inexpensive glassware bought from a discount online retailer, and to be honest, this isn't always a terrible idea. There is certainly a context where the non-technical solution is an acceptable, even a desirable one. Where possible,

DOI: 10.4324/9781003391708-2

this is indicated within the relevant recipes. The aim of this section of the text is to help the conservator recognize which compromises are being made, not to dictate 'best' methods.

All that said, if purchasing from non-scientific suppliers, do not expect the materials to behave in the same way, and to the same standards as high-quality laboratory equipment. Equipment listed on popular online retailers frequently either lacks tolerance or interval information or is labelled incorrectly, even misleadingly. This is an issue for 'dupes', cheap versions of established brands, as well as actual counterfeit items, both of which can be easily found on popular dropshipping websites [2].

How much this matters will depend on the equipment in question and the degree to which reliability is important. For example, if a procedure calls for heating or cooling a sample, working with heattolerant glassware is essential as non-heat-tolerant glassware may crack or break during the procedure, creating a hazard as well as leading to a loss of material. There is also one final ethical point to make. Supply chain opacity supports illegal, immoral, and reprehensible production systems like trafficking in conflict materials, forced labour, and slavery [3, 4]. Now, despite the severity of the subject, is every bit of plastic sold online the result of these practices? No, almost certainly not. The trouble is that supply opacity leaves the system open to such abuses. While far from perfect, it's worth noting that suppliers based in the UK and EU are now required to address these concerns and are barred from such practices [5, 6]. Even the act of writing this feels heavy-handed, but it's worth being aware of the technical and ethical challenges so often posed by the more economic options.

## Accuracy

Consultant: Thomas Higgins

Rather than starting things off with something simple and intuitive, we will instead first look at accuracy. I'd argue that for certain instances in practical conservation work, some imprecision can be tolerated. A level of ambiguity and adaptability is built into the space between solution preparation and the point of application, as materials are routinely altered either by intentional actions or passive processes. Many conservators no doubt allow a bit of solvent to evaporate off the brush before applying a slightly too-thin adhesive or will prewet the brush if a consolidant is just a little too thick. In these instances, an estimated percent composition is recorded for documentation but isn't always a true reflection of the exact amounts of materials used.

And many times, this level of accuracy is fine. Until, that is, it's not. Again, the most important skill is not necessarily found in eliminating any ambiguity from measurement, but is instead rooted in clearly understanding the compromises being made.

In the brush-level examples previously described, this variation is the informed choice of the individual conservator, not a side effect of inappropriate preparation techniques or inaccurate equipment. Conversely, if a skilled conservator was using an inaccurate balance to prepare an adhesive, the resulting solution could be adapted prior to use. While less than ideal, here the likely effect of inaccuracy is simply incorrect documentation. But if that same balance was used to prepare a chelating solution, resulting in a 5% EDTA concentration rather than the intended 2%, the results could be catastrophic.

This section is intended to make something I personally find a bit challenging into something both intelligible and actionable. These ideas will therefore come up repeatedly in the following sections, where they have been adapted to specific considerations for each type of measuring equipment. In order to facilitate better understanding of the particular challenges, some terms will need to be defined.

### *Key Terms*

**Accuracy** – How close a measurement is to the true or accepted value of the measured quantity. If the true value is known, accuracy can be measured by percent error.

**Precision** – How tightly grouped a series of measurements are. Precision can be measured by calculating the range or the sample standard deviation of a set of measurements.

The classic illustration of this idea uses a dart board. Note how the accurate thrower hits near to the centre, to the 'true' value, but maybe not too close to each other. The precise thrower keeps their darts tightly grouped, but they may be considerably farther from the bullseye. A master dart thrower will be both accurate and precise, hitting the centre with all their darts. However, this does not imply that the two ideas are mutually exclusive; they are not. For just as I can throw a bullseye despite being utterly useless at darts, accuracy without precision is just luck (Figure 1.1).

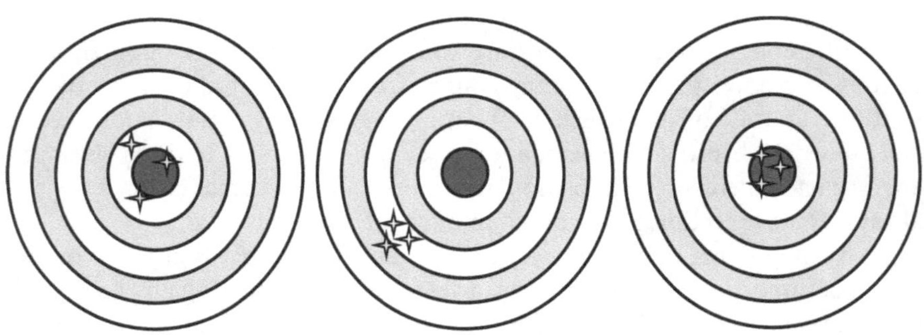

*Figure 1.1* Illustration of accuracy and precision using a dart board as a metaphor.

**Significant figures** – A method of determining that correct number of figures which will most accurately reflect the uncertainty of the reported value. Practically, it is used to determine when to round off a figure, or how many decimal points to include. See the longer section for further details and examples.

**Tolerance** – Tolerance is a manufacturer-supplied value that essentially shows the amount of uncertainty or deviation considered acceptable for a specific item of measuring equipment. This can also be conceptualized by the acceptable level of inaccuracy. For example, a 250 mL graduated cylinder with a tolerance of 0.8 mL means that any reading from that cylinder can be in actually nearly 1 mL less or more than the stated amount. This is most commonly used in volumetric measures.

**Readability** – Used in electronic balances, this describes the number of digits that are displayed but should not be confused with the accuracy of the device.

**Range and intervals** – All measurement equipment operates within a finite, predetermined range of values. Although that range will vary from item to item, no bit of benchtop kit can measure infinitely small and infinitely large quantities with the same measure. The upper and lower limits of this are the scale minimum and maximum. This range is then further broken up into a series of smaller increments, the ticks on a graduated cylinder or the zeroes on a balance, and these define the smallest measurable amount or the interval (often listed as *d*).

For example, a 250 mL graduated cylinder with markings every 2 mL and a tolerance of 0.8 mL has a range of 2-250 mL and an interval of 2 mL. This cylinder can therefore only measure to even-numbered mL with an accuracy of ±0.8 mL.

***Significant Figures***

Adapted from T. Higgins (2024) [7]

As mentioned in the short definition given previously, the 'significance' of a digit within a larger number is determined by applying a series of (mostly) easily applied rules. This provides a consistent method for reading and recording numbers during measurements and calculations.

At a practical level, you will save yourself and whoever tries to copy your work a great deal of time if you adhere to the following practices when reading and recording data. These are by no means perfect, but they represent a reasonable trade-off between time and accuracy. If you encounter a situation where you absolutely need to know the number of significant figures in a measurement or a calculation for publication or such, you will probably be better off using error analysis software with the help of a colleague.

Finally, despite its notational proximity, 'significance' in this context does not have anything to do with importance. Just because a number isn't significant doesn't mean it can just be removed. This will become apparent in discussions of 'leading zeros' but is worth explicitly noting here as well.

*Rounding*

Rather than addressing this topic in depth, here are two pragmatic and simple practices that can reduce anxiety about significant figures and calculations.

1. When recording data always record every digit and never round. If the balance says "'2.00"', write down "'2.00"', not 2. If your data always has the right number of significant digits, someone can always double-check your calculations.
2. When reporting data, only round figures at the very end of calculations.
3. When doing calculations, always record four significant digits or two digits past the decimal – whichever is more. If you record too many digits, it is easy to round a number *down*. If you record too few, you have to redo the calculation, which may be time-consuming.
4. When rounding, use only the digit one to the right of the significant value. For example: Rounding 2.362 to three sig figs gives a final value of 2.36, while rounding to only two sig figs gives 2.4.

*Identifying Significance*

Significant figures are simply those that lie between the first non-zero number on the left of a value and the last number on the right. This second criterion is what causes trouble, as the significance of the 'trailing' zeroes on the right can be ambiguous, especially if the number does not have a decimal place.

For example: according to these rules the number 1.00 unambiguously has three significant figures. But the number 100 could have one or three. There is no way to know. Absent any other information, this author suggests treating it as having three.

There's also the counter-intuitive issue of the 'leading' zeros. For some reason a figure like 0.0000025 feels more significant than a value like 825, but if you remember your definition about only counting the first non-zero number from the left, 0.0000025 has only two significant figures, while 825 has three. Again, this isn't to say that all those leading zeroes aren't important, simply that they are not *technically significant*.

Using the number 0.0050600 kg as an example, let us discuss a few key ideas.

- Leading zeros – As the name implies, these are the zeros that precede a number. These are *never significant* but must be included to preserve the value of the number.
- The zero in the middle, between the "'5"' and the "'6"', is called a *captured zero*. Captured zeros are *always significant*. Removing them would change the value of the number.

- The two zeros on the end, after the "'6'", are called *trailing zeros*. These may or may not be significant and their significance cannot be determined without further information. It is always best to consider trailing zeros significant unless there is a good reason not to.

A few further examples:

| | |
|---|---|
| 2.00024 | 6 sig figs |
| 0.003 | 1 sig fig |
| 0.300 | 3 sig figs. Remember, trailing zeros to the right of the decimal count as significant. |
| 100.0 | 4 sig figs |
| 100 | Hard to tell since trailing zeros without a decimal are ambiguous. Assume three. |

*Applying Significance*

This might seem somewhat abstracted from practical work, or even a bit elementary, but it has direct bearing on accuracy. It can be helpful to remember that significance essentially keeps track of the numbers that are known knowns. If a weight is given as 110 g, say, I cannot treat it as if it were 110.00, because I can't say for sure if that's true. Rounding correctly helps to keep track of how many digits are, in fact, known.

***The Maths***

Here are a few rules on how to keep the correct number of significant figures through various simple operations. There is also extensive information on chemistry tutorial websites should this section leave specific questions unanswered.

Please keep in mind that when working to correct significant figures, do not round until the calculations are complete; instead keep all figures and round only at the very end. Furthermore, round using only one position to the right, and ignore any further figures.

*Adding or Subtracting*

Simply round to the smallest decimal value found in all figures in the set.

**Example 1**

The sum of series: 110, 15.2, 3.26

As 110 does not have a decimal value, the sum must be rounded to the nearest whole number. This can be easier to see if you imaging the values are stacked vertically

$$
\begin{array}{r}
110 \\
15.2 \\
+3.26 \\
\hline
128.46 \\
=128
\end{array}
$$

Note that only the numeral immediately to the right is used for rounding, the 6 in this instance is not significant and is therefore not included.

**Example 2**

The difference between 100.4, 4.04, 35.98

The smallest decimal shared by all three numbers is the tenths value (0.1)

$$100.4 - 4.04 - 35.98 = 60.38$$
$$= 60.4$$

*Multiplying and Dividing*

When multiplying or dividing, count the number of significant figures for each number in the series. The number with the least sig figs sets the correct value for the final number. As with adding, the final value can only be known to the level of the least significant part.

**Example 1**

The product of series: 11, 15.2, 3.26

11 has only two sig figs, 15.2 has three as does 3.26. The final answer should therefore only be rounded to two significant figures.

$$11 \times 15.2 \times 3.26 = 5450.72$$
$$= 5500$$

As with adding and subtracting, it is only the numeral immediately to the right that is used for rounding. This can sometimes result in answers that feel wrong, that seem not to accurately reflect the component parts. Again, however, your answer can reflect only what is known for your least significant value; it's only as good as the 'worst' measurement.

**Example 2**

The quotient of 100.4, 4.04, 35.98

100.4 has four sig figs, 15.2 has three, as does 3.26. The final answer should therefore be rounded to three significant figures.

$$100.4 \div 4.04 \div 35.98 = 0.690702755\ldots$$
$$= 0.691$$

Mixed operations. Keep the same rules, selecting the least sig figs found within the larger equation. For example:

$$(7.01 \times 1.235) + 2.000 + (3.41/3.0) + 4.52 = 16.314016\ldots$$

The figure with the smallest number of significant figures for the operation is 3.0. The answer therefore has only 2 sig figs and should be rounded to 16.

One final caveat: there is also the idea of 'exact numbers'. In this context, it refers to values which are likely to appear in conversion factors. For example, 50°C equals 122°F; if used as a conversion factor, these numbers should not be rounded. Technically such numbers are considered to have infinite sig figs and are therefore ignored. Objects which can be counted are also considered exact. There is no uncertainty about the number of fingers on your hand.

Take, for example a molarity calculation like those discussed in Calculations Section.

$$\frac{1\,\cancel{mol}}{L} \times \frac{98.079\,g}{\cancel{mol}} = 98.079\,\mathrm{g/L}$$

Note that the conversion factor, the mass of sulphuric acid ($H_2SO_4$) per mole does not cause the sig figs to drop to only one. Instead, the "per mole" is considered an exact number, and its significance is therefore ignored in these calculations.

### *Uncertainty*

Finally, we take a slightly more in-depth look at the practical aspects of uncertainty. Because of the very small batches of very low concentration mixtures that are common in conservation, there is a temptation to work at the very edges of accuracy. For example, using a balance that measures down to a thousandth of a gram in decimals (0.000) to measure out three milligrams (0.003 g) of materials. Unfortunately, as is true for all for all such equipment, the closer the value is to the minimum that can be measured, the greater the effect of uncertainty.

This is the result of simple ratios. For example, an uncertainty of ±0.1 g for a 50 g sample works out to around 0.2% possible error for the total weight, but the same uncertainty for a 2 g sample is 5%. This means that small solutions will be inherently less accurate than larger batches. How inaccurate will depend on your balance, the local conditions, and when, if ever, it was last calibrated.

The balance from the previous example with an uncertainty of ±0.1 g will give readings that are most likely to sit somewhere within 0.1 g more or less than the true value. A known weight of 50 g on such a balance will therefore probably give readings anywhere between 49.9 g and 50.1 g.

Please note that this is a pretty gross oversimplification of a very important topic that is only intended to point to the tip of the iceberg and clarify why these things matter. If you're interested in how this can cause confusion in scientific reporting, and you have journal access, please see Landström et al. (2015) for an excellent discussion of how uncertainty is communicated across different disciplines [8].

## Measurement

Consultant: Thomas Higgins

In order to avoid burying the lede (even further) in paragraphs of technical information, each measurement type will open with specific recommendations or common mistakes before providing context and rationale. However, the technical information contains the bulk of the information required for good understanding and it is therefore suggested that you read on for greater context.

### *Volume*

#### *Starting Rules*

1. Temperature affects volume. All materials shrink and swell as the temperature cools and heats. 10 mL of material at 0°C does not have the same amount as 10 mL of material at 25°C.
2. Volume is not a conserved quantity like mass is. When mixing liquids 1 + 1 ≠ 2. To prove this to yourself, mix 100 mL of water with 100 mL of rubbing alcohol and measure the volume of the mixture.
3. Whenever possible, measure and record the mass of a material. Mass does not change if temperature changes. Mass does not change when materials mix.

#### *Dos and Don'ts*

1. First and foremost, as a rule, don't use beakers for measurement.
2. Opt for the smallest measurement system which can contain the full volume of the liquid being measured. Note that measuring a liquid is not the same as storing a liquid. For safety reasons, try not to fill a container (like a

beaker) more than half full with a liquid. Overfilling makes it more likely a liquid will spill, potentially creating a hazard.
3. Measure only to graduation marks and within the tolerances of the specific glassware.
4. Very small volumes should be measured by glass pipette or dispensing micropipette. Disposable transfer pipettes are not designed as measuring equipment. Small volumes can sometimes be accurately measured by mass if the density is known.
5. Scoops and cups used for home cooking can be used for volume:volume ratio mixtures (1 tbsp added to 1 tbsp), but are not accurate enough for volumetric measurements

*Beakers*

Many beakers have only very approximate volume markings rather than true graduations and are intended only to show how much they can hold. They are therefore best used for mixing rather than measuring. For safety, and to limit spills, avoid filling a beaker more than half full when possible.

*Graduated Cylinder*

This is perhaps the most common volumetric measure found in conservation spaces. They're simple to use, long-lasting, inexpensive, and accurate enough for most treatments.

The 'graduations' of the cylinder are the small tick marks that run up the side. In small volume cylinders, these are usually around every 0.5 mL, while in larger ones it can be up to 2 mL. The volume between these ticks represents the smallest possible measure and is therefore related to, but not the same as, the tolerance of the cylinder. Although this information is rarely written on the cylinder, a broad rule of thumb seems to be that maximum tolerance is often around one-quarter to one-half the graduation [9].

The reason to discuss tolerance is that it simply provides the technical reason why smaller amounts are best measured in smaller cylinders. To clarify this, one needs only compare two hypothetical scenarios: measuring 5 mL in a 10 mL and 20 mL cylinders respectively.

In the 10 mL cylinder, the graduations are every 0.5 mL and the tolerance is roughly ±0.25 mL. The 5 mL can be measured directly against a graduation mark, and the final true volume is therefore between 4.75 mL and 5.25 mL. This results in an error of around 5% of the total volume.

In the 20 mL cylinder the graduations are every 1 mL and the tolerance is roughly ±0.5 mL. The 5 mL can be measured directly against a graduation mark, and the final true volume is therefore between 4.5 mL and 5.5 mL. This results in an error of around 10% of the total volume.

In this hypothetical situation, the doubling of the cylinder volume also doubled the percent error of the resultant solution. This demonstrates that even when you are able to measure directly against a graduation mark, it is always more accurate to use the smallest possible cylinder that can still hold the entirety of the volume being measured.

That said, although smaller can be more accurate, do not simply take two measurements then add them together. Multiple measurements compound their errors and any benefits from using the lower tolerance cylinder are immediately completely lost.

A short note about markings. Until recently, the UK designated between two types of graduated cylinder, the more precise A, and B, with the latter having twice the tolerance of the former. The type would be clearly inscribed on the cylinder, and while this can be found on older glassware or those sourced from other regions, this system is no longer in use in the UK [10, 11]. Glassware is also often inscribed with the ideal temperature for accurate measurement, typically around room temperature (20°C/70°F).

### *Small Volumes*

Micropipettes, such as those most often encountered in the biochemical context, are a very cost-effective approach for measuring (and dispensing) small quantities of liquid. The exact accuracy and tolerance will depend on the unit chosen, but they regularly range between 2 µL to 1000 µL (1 mL) in single microlitre intervals. The operation is more controlled than disposable plastic pipettes, and they dispense more smoothly than syringes. The tips are small and easy to wash, and if disposal is required, the amount of plastic ending up in the bin is much less.

It's also worth noting that 'measurement' doesn't appear under the many uses for the regular plastic disposable transfer pipettes [12, 13]. It seems that like those for beakers, the markings on the sides of transfer pipettes are for rough approximations only.

### *Scoops and Cups*

Volumetric scoops and cups such as those commonly used in cooking are particularly helpful in v:v preparations that support a bit of inaccuracy. Wheat starch pastes and further dilutions of spreadable gels are two easy examples where scoops are especially useful. In these volumetric ratios, the specific unit doesn't matter, only that it is consistently applied. A dilution of 1 tbsp of 2% xanthan gel with the addition of 1 tbsp solvent will produce the same concentration as a 20 mL: 20 mL dilution of the same. It's the ratio that is important in this context, and not the units.

However, a quick note that although domestic measures like cooking cups and teaspoons are perfectly acceptable in the context of v:v ratios, tools designed for home use are simply not subject to the same standards as those designed for the laboratory. As such, they can be incompatible with solvents and wildly inaccurate, with non-comprehensive reports of up to 30% variance in scoops/spoons and 13% in measuring cups [14].

### *Mass/Weight*

*Dos and Don'ts*

1. Don't rely on the final digit on the display. Instead, use the final 1 or 2 digits only as approximate guidance. The number of digits displayed on a balance is not necessarily correlated with the smallest value it can reliably measure.
2. Always record the full measurement, but check the tolerance of the balance when making sig fig determinations.
3. Regularly calibrate. What is required for calibration will depend on the specific balance. In a pinch, however, you can check calibration with a minted coin. The mass of these is typically standardised to ± 0.02 g. For example, a 20 pence piece with QEII on the front and a US 5¢ nickel both weigh 5.0 g and the 20 cent Euro is 4.10 g. Note that the coin must be clean, dry, and not too worn down.
4. Don't move an analytical balance once it's been calibrated. At all. Yes, even sliding it down the bench. Some precision balances can be moved, notably those that are advertised as 'portable' or 'stackable', but unless this is explicitly stated, the assumption should be that movement is to be avoided.
5. If possible keep it plugged in to mains power. While not true for all, many electronic balances must be kept powered at all times, even when not in use. These can require either constant electric current or a minimum of one hour 'warm up' prior to taking any measurements.
6. Avoid working too near either end of the provided measurement range. Amounts approaching the minimum are more likely to be inaccurate and amounts above the maximum can damage the equipment.
7. Keep the weight of the measuring container relatively similar to the amount being measured.

*Electronic Balance*

There's really only one tool used in the preparation of conservation solutions by weight, the tabletop electronic balance.

There are only two main types of electronic balances. The electronic precision balance (EPB) typically reads up to a maximum of three decimal places (0.000) and is not really affected by temperature or draughts. For this reason, EPBs rarely have draught shields or other encapsulating features. Then there is the electronic analytical balance (EAB), which reads over 4 decimals and is designed for precise weighing. EABs are often used in standardized formulation or government-regulated preparations. These typically do have a draught shield as they are affected by temperature, humidity, and vibration [15].

The best guidelines for the use and care of a given balance will be those supplied directly from the manufacturer. If this brochure has been lost, I strongly recommend trying to chase down a copy through the manufacturer's 'website or by reaching out to the supplier. Check the documentation to ensure it matches the model number of your balance and isn't just something similar. The information discussed below reflects some general best practice, but individual results may vary.

Although the electronic balance is extremely common, in my experience it is remarkably easy to misread, mishandle, and misinterpret. This is a completely transparent attempt to hook you, the extremely experienced professional, into at least skimming the following section.

### *Readability and Range*

Accuracy is often conflated with readability. While there can be a relationship between the two, it is not causal, and more numbers do not mean higher accuracy. A precision balance that reads to four figures (0.000 g) is not inherently less accurate than an analytical balance that reads to 6 figures (0.00000 g). Instead, the first is likely intended to measure within the range of tenths of a gram, whereas the latter is designed for milligrams. It is a question of scale, and not of accuracy [15, 16].

This distinction is why every balance has a given *range* with a minimum (usually 0) and a maximum divvied up by a scale *interval* of *d*. Please note that, as with graduated cylinders, the interval marks the smallest distinction that can be made between measures. Therefore, values less than *d* cannot be accurately measured.

Further confusing things, although *d* is the smallest increment for a given balance, this value does not relate to the smallest *amount* that a scale can reliably measure, which is referred to as the minimum weight. To reiterate, the minimum weight is the smallest amount that a balance can reliably measure, a value that is greater than zero and unrelated to how many figures are displayed [17]. There is also a maximum load, above which readings can be similarly inaccurate, but when exceeded runs the added risk of damaging the components in the balance itself [18].

The load maximum and the interval (*d*) are usually printed directly on the balance somewhere near the display, whereas the minimum weight must come from the manufacturer.

All this context is to say that the accuracy of a balance is not a result of how many digits it displays, nor how fine the intervals it can measure. Accuracy is determined by the build of the balance and local conditions the combination of which is best described by calibration.

### *Calibration and Conditions*

In order for a balance to provide accurate readings, it must be regularly calibrated. This is particularly important for higher precision instruments, for as the sensitivity of the balance goes up, so too does the potential influence of external factors such as temperature changes and vibrations. This shift in accuracy, sometimes cumulative, but other times finite, is known as drift [16].

Calibration not only brings an electronic balance back to more precise and accurate readings, it does so in direct response to the *specific conditions* the balance is placed within. This means that a balance calibrated for one location will not give the same precision or accuracy if *any* of the conditions change [19, 20].

This hyper-specificity demonstrates why you really shouldn't be moving an electronic balance around – particularly if you're taking the time and spending the money bothering to calibrate your equipment.

Similarly, due to the sensitivity of components, excess weight too strongly applied can also damage delicate measurement equipment. For example, a large glass beaker plonked directed on an analytical balance is both too large for the equipment and was applied too vigorously. If measuring small amounts, use lightweight weighing boats or paper rather than heavy watch glasses or other glassware [18].

Finally, electronic balances must also be 'warmed up', supplied with consistent power for 30 minutes to one hour prior to measurement. This ensures that the delicate electronic components reach the desired thermal equilibrium required for precise readings [18].

Again, however, how much precision is required will depend entirely on the objects being treated and the solutions being prepared. More subjective workability-based measures, such as w:v adhesive mixtures, can often tolerate less precise measurement.

### *pH*

For studios and laboratories in need of consistent, reliable, precise, and accurate pH adjustment, a handheld pH meter is a worthwhile investment. These typically require calibration prior to each use, but the process is relatively painless, with many meters supplying the required buffer solutions. Also, given that the pH scale is logarithmic, it is worth it to spend a bit more to get a pH meter that reads to two places past the decimal rather than just one. However, such high sensitivity equipment requires regular checks and a stable environment as improper storage and fluctuations can affect accuracy. Finally, these handheld meters work very well with liquids, but have shown some inconsistencies when working with synthetic gels like Pemulen and Carbopol [21].

If infrequent testing is needed or a low level of accuracy is acceptable, pH indicator strips can also work well. These are subjective and relatively imprecise, with accuracy ranging from 0.3 to 1 pH. That said, they are cost-effective, and strips with multiple indicator windows have been found to give consistent results [22]. These strips rely on reagents, which do lose their potency over time, so expiry dates should be checked prior to use.

## Calculations

Consultant: Thomas Higgins

The following section is designed to provide a quick reference to calculations which are only sporadically encountered in conservation treatments. The information is largely provided without too much context with the exception of the following short discussion on concentration.

### *Concentration*

This is receiving special attention simply because it seems to mark an issue of inconsistency among conservation documentation. It is worth a quick reminder that the solute is the substance being dissolved while the solvent is the substance doing the dissolving. The combination of the two is the solution. Keep in mind that the solution has chemical and physical properties that are distinct from each of the components and should be considered a distinct material. These relational descriptors are independent of the phase of the materials, be they solid, liquid, or gas.

#### *Weight and Volume*

##### NON-CONSERVATION OF VOLUME

As mentioned above, volume is not a conserved quantity, so mixing 10 mL of solute with 10 mL of solvent does not guarantee 20 mL of solution. This is why it is important to use volumetric glassware when preparing solutions true w/v solutions. It is also why you can't just calculate the volume taken up by a solute using the density. As soon as something dissolves, its density is no longer the same.

##### CONCENTRATIONS AND RATIOS

Perhaps the most common, and also the most commonly mistaken, concentration value is the weight-to-volume or w/v. This concentration gives the weight of the *solute* against the total volume of the *entire mixture*, solute and solvent combined.

A 40% (w/v) mixture of Paraloid B72 is not 40 g of B72 in 100 mL of solvent, but instead 40 g of B72 in approximately 65 mL of solvent. This latter value is the correct one, because it takes into account the volume of the B72 beads, which, especially at 40%, is a significant value.

Now, you will absolutely encounter circumstances where concentrations are so low as to make such distinctions moot; 0.5% concentrations of methyl cellulose are a good example. Half a gram of methyl cellulose takes up such a small amount of space when compared against the roughly 100 mL portion of water; you're certainly working outside practical significance with the amounts given, and possibly working outside the accuracy of your equipment.

Furthermore, sometimes it's also simply not terribly practical to make true w/v concentrations. And that's fine. As discussed previously, with non-reactive mixtures like adhesives, if it has the desired properties, then it doesn't necessarily matter if it's a true w/v. However, in these instances the conservator must appreciate the distinction, and the documentation should reflect it.

This text therefore adopts the convention of referring to true w/v concentrations as such, while ratio mixtures are instead listed as w:v. This is not without its flaws, but it is hoped that *any* differentiation will help bring increased clarity and consistency to understanding and documentation.

### CONSERVATION OF MASS

Much of this confusion can be completely sidestepped by the widespread adoption of weight/weight concentrations. As I'm yet unaware of any conservation laboratories on the moon or elsewhere not on Earth, the terms *mass* and *weight* will be considered analogous. Unlike volume, mass is a conserved property, so no additional steps or special glassware are required.

Although the density of any other solvents will differ, with water, it's worth remembering that at 10°C 100 mL of water weighs very nearly 10 g. To prepare a w/w solution, simply weigh out the mass of solute, and subtract that from the directed final solution weight to get the amount of solvent needed. Then weigh out both parts separately and add the two together. Because it is conserved, weight, unlike volume, is additive, and the total will equal the sum of the parts. Another approach is to measure the weight of the solute, and simply add solvent until the final desired weight is met. Due to the simplicity of this approach, the text gives strong preference for w/w concentrations where it is easily applied.

Unfortunately, it isn't often easily applied. Much conservation literature uses the w/v standard, and many people, myself included, are accustomed to thinking in terms of volumetric concentrations. A 40% (w/v) solution of Paraloid B72 has a consistency that many of you are no doubt picturing as you read this. To make matters worse, because volume isn't conserved, and it's therefore not a simple matter of density calculation, moving between the two systems is not straightforward. A researcher with a bit of time and access to high-quality measuring equipment could certainly devise a reference table to help bridge the gap (nudge, nudge), but this is outside the scope of this text (and capabilities of this researcher).

### EXAMPLE MATHS

*Volumetric Concentration (w/v) Example* Prepare 10 mL of exactly 20% weight/volume solution of Aquazol 200 in water

Volumetric container

1. Calculate how much solute is required.

$$\frac{10\ \cancel{mL}\ solution}{} \times \frac{20\ g}{100\ \cancel{mL}} = 2.0\ g\ solute$$

2. Weigh out 2.0 grams of Aquazol and add to a volumetric container (a wide-mouth graduated cylinder or volumetric flask).
3. Add some water to the volumetric container and stir. Allow the Aquazol to fully dissolve.
4. Once dissolution is complete, add more water until the level meets the 10 mL mark.

*Volumetric Ratio (w:v) Example* Prepare 10 mL of a 20:100 weight:volume solution of Aquazol 200 in water

1. Calculate how much solute is required.

$$\frac{10\,\cancel{mL}\ solvent}{} \times \frac{20\,g}{100\,\cancel{mL}} = 2.0\ g\ solute$$

2. Weigh out 2.0 grams of Aquazol and measure out 10 mL of water.
3. Add both to same container and mix until dissolved.

*Weight Concentration (w/w) Example* Prepare 10 g of exactly 20% weight/weight solution of Aquazol 200 in water

1. Calculate how much solute is required.

$$20\% \times 10\,\text{g} = 2.0\,\text{g solute}$$

2. Place the lidded container on the balance and press *tare*/zero
3. Add Aquazol pellets until you reach 2.0 grams, immediately add water (without taring), and top up until you reach 10 g.
   *or*
   Weigh out exactly 2.0 g of Aquazol, and then weigh out exactly 8.0 g of water. Add both to the container
4. Close the container and shake until well dissolved.

DISCUSSION

These three Aquazol solutions will have different working properties due to the variations in the amount of resin and solvent present. In the first example (w/v concentration), there are 2 g of solute with roughly 7–8 mL of solvent; this is the thickest solution. In the second (w:v ratio), there are 2 g of solute with 10 mL of solvent; this is thinner than the first solution. The third example (w/w concentration) has 2 g of solute and 8 g of solvent; remember, however, that 8 g of water is around 8 mL. This last example is therefore far more dilute than the previous two.

*Table 1.1* Approximate values for Paraloid B72 and acetone.

| *Mix type* | *Parts* | *Parts* | *w/v Concentration* |
|---|---|---|---|
| w/v | 0.5 / 100 | 0.5 g into 99.6 mL | 0.5% |
| | 1 / 10 | 1.0 g into 9.2 mL | 10% |
| | 5 / 10 | 5.0 g into 5.6 mL | 50% |
| w:v | 0.5: 100 | 0.5 g into 100 mL | 0.5% |
| | 1: 10 | 1.0 g into 10 mL | 9.2% |
| | 5: 10 | 5.0 g into 10 mL | 35% |

These are illustrative of the difference between the two solution types and *do not represent* actual known values. As discussed previously, volume is not conserved, so the dissolved B72 will take a up a different volume from that implied by its density as a solid.

### *Molarity*

#### TERMS

Moles (mol) – the unit used to compare the number of atoms or molecules in balanced chemical equations.

Molarity (M) – number of moles of solute per one litre of solution.

Molar mass –the mass of exactly one mole of a given element, molecule, or compound. In this context it is used as a conversion factor for changing moles into grams. This is often expressed in the unit of atomic mass unit (u) but this is numerically the same as grams per mole (g/mol)

#### MATHS

Prepare solution to desired molarity.

1. Determine the molecular formula of the solute.
2. Often the molar mass of a substance can be found in documentation like a safety data sheet (SDS) or looked up from a reputable source like PubMed. When possible, this is preferable since those values have been error-checked.
3. Use a periodic table to look up the atomic weight for each element in the formula. Add up the atomic weights of each element to determine the molar mass.
   For example:

$$\text{Methanol}: CH_3OH = CH_4O = 1\,\text{carbon} + 4\,\text{hydrogen} + 1\,\text{oxygen}$$
$$1\,(12.01) + 4\,(1.01) + 1\,(16.00) = 32.05\ \text{g/mol}$$

$$\text{Ethanol}\ CH_3CH_2OH = C_2H_6O = 2\,\text{carbon} + 6\,\text{hydrogen} + 1\,\text{oxygen}$$
$$2\,(12.01) + 6\,(1.01) + 1\,(16.00) = 46.07\ \text{g/mol}$$

4. Convert from desired final concentration to grams using molar mass as a conversion factor

$$M = \text{mol} / \text{litre}$$
$$\text{mol} / \text{litre} \times \text{g} / \text{mol} = \text{g} / \text{litre}$$

EXAMPLES

Prepare a 10 M solution of sodium hydroxide.

1. Determine the formula of sodium hydroxide.
   NaOH
2. Calculate the mass of one mole of sodium hydroxide. Add up the atomic weights of the amount of each element.

$$1\,\text{sodium} + 1\,\text{oxygen} + 1\,\text{hydrogen}$$
$$22.99 + 16.00 + 1.01 = 40.00\,\text{g} / \text{mol of NaOH}$$

3. Convert from desired final concentration (10 M here) to grams using molar mass as a conversion factor.

$$10\,M = \frac{10\,\text{mol}}{L}$$
$$\frac{10\,\cancel{\text{mol}}}{L} \times \frac{40\,\text{g}}{\cancel{\text{mol}}} = 400\,\text{g} / \text{L}$$

4. Mix: mostly fill a 1 L volumetric container with water, slowly add the salts and then top up the remaining water.
   Safety note: When mixing large volumes of materials (>100 mL) do so slowly. Dissolution of salts like sodium hydroxide liberate a lot of heat, and if done too quickly, can cause splatter of the solution and lead to chemical burns. The heat can also damage or crack non-thermal glassware.

Prepare a 100 mL solution of 3 M citric acid in water

1. Determine the formula of citric acid

$$HOC(CO_2H)(CH_2CO_2H)_2 = C_6H_8O_7$$

2. Calculate the mass of one mole citric acid. Add up the atomic weights of the amount of each element

$$6\,\text{carbon} + 8\,\text{hydrogen} + 7\,\text{oxygen}$$
$$6(12.01) + 8(1.01) + 7(16.00) = 192.14\,\text{g} / \text{mol of citric acid}$$

3. Convert from desired final concentration (3 M here) to grams using molar mass as a conversion factor

$$3\,M = \frac{3\,\text{mol}}{L}$$

$$\frac{3\,\cancel{\text{mol}}}{L} \times \frac{192.14\,g}{\cancel{\text{mol}}} = 576.42\,\text{g}/\text{L}$$

4. Convert from grams per litre to grams per 100 mL

$$10\,M = \frac{10\,\text{mol}}{L}$$

$$\frac{576.42\,\text{g}}{\cancel{L}} \times \frac{1\,\cancel{L}}{1000\,\text{mL}} = 0.57642\frac{\text{g}}{\text{mL}} \times 100\,\text{mL} = 60\,\text{g in}\,100\,\text{mL}$$

5. Remember, although citric acid is mild, water should never be added to acids; instead acids should ways be added to water. As a result, this solution requires an additional step. Pour around half the final volume of water (50 mL in this case) into a beaker. Add the 58 g of acid to the water and stir until dissolved.
6. Pour the full contents of the beaker into the volumetric container.
7. Rinse the beaker 2–3 times with approximately 10 mL of water to ensure all of the citric acid solution has been transferred.
8. Fill with water until you reach the desired final volume (100 mL).

RATIONALE

Weight-to-volume concentrations are most helpful when preparing solutions with industrially protected formulas (such as many adhesives) or extremely large and complex molecules (like enzymes). However, when preparing solutions designed to produce specific chemical reactions, molarity is a much simpler and more precise way to calculate concentration.

Using molarity allows us to translate the reactions described in chemical equations directly into practical amounts to be mixed. Although some of the terms can seem unfamiliar to the less chemically confident among us, calculating molar concentration is no more challenging than the other methods outlined previously and is considerably more precise. The main difference is that rather than comparing weight to volume, and therefore muddling into questions of density and molecule size, you're directly comparing the number of molecules. This is necessary because a 5% w/v concentration of a small salt like NaCl in water will have many, many more molecules in solution than a 5% w/v concentration of a larger molecule like methyl cellulose.

Moles give a standard system that lets us compare the number of various molecules, atoms, and salts in a manner that is independent from differences in mass or volume but reflects roughly how they will chemically react. This can seem a bit abstract until you realise that balanced chemical equations describe reactions between the numbers, or moles, of molecules.

$$2\,H_2 + O_2 \rightarrow 2\,H_2O$$

Two moles of hydrogen mix with one mole of oxygen to produce two moles of water. You can now see how this is helpful. If you were trying to create water, in some cursed laboratory, from hydrogen and oxygen, you would need two mole equivalents of hydrogen for every one mole equivalent of oxygen for all elements to be converted to water without any excess.

Now that moles are suitably understood, molarity simply an expression of molar concentration, or the number of moles of solute into exactly 1 litre of solvent. A one molar (1 M) aqueous solution of NaOH has 1 mole of NaOH per litre of solution, a 5 M solution of the same has 5 moles of NaOH per litre of solution.

Molarity is also immensely helpful in planning dilutions.

### *Dilution*

#### *Equations and Terms*

$$\mathbf{C_1 \times V_1 = C_2 \times V_2}$$

$\mathbf{C_1}$ = starting concentration (the stock solution to be diluted)
$\mathbf{V_1}$ = volume of stock solution required to prepare new mixture (usually the unknown)
$\mathbf{C_2}$= desired concentration of the new solution
$\mathbf{V_2}$= desired final volume of new solution.

Concentrations most often given in molarity (M) or percents (0.1 or 10%). Volume can be given in any unit measure. No matter the measurement used, the units on both sides of the equation must be the same.

#### *Maths: Example*

Prepare 5% concentration of adhesive in acetone from a 20% stock solution of the same. First, approximate how many mL of solution you want for the end product. In this example, the adhesive is only for a few small repairs so 30 mL would be more than enough. In this instance:

$\mathbf{C_1}$ = stock concentration = 20%
$\mathbf{V_1}$ = volume of stock solution required to prepare new mixture = unknown
$\mathbf{C_2}$= desired concentration of the new solution = 5%
$\mathbf{V_2}$= desired final volume of new solution 30 mL

$$C_1 \times V_1 = C_2 \times V_2$$
$$20\% \times V_1 = 5\% \times 30\,\text{mL}$$
$$0.20 \times V_1 = 1.5$$
$$V_1 = 7.5\,\text{mL}$$

Finally, the 7.5 mL of stock solution must then be diluted to the desired 30 mL with additional solvent.

Again, to get an accurate solution, use of a volumetric measure is required. However, if approximate values are acceptable, you can determine roughly how much more solvent is needed by simply subtracting the amount of stock solution just calculated ($V_1$) from the desired final volume.

$$30\,\text{mL} - 7.5\,\text{mL} = 22.5\,\text{mL}$$

Therefore: to make up 30 mL of 5% (w/v) mixture from a 50% (w/v) stock solution, decant 7.5 mL of the stock solution to a container and either top up to reach 30 mL or, for an approximate value, add 22.5 mL of additional solvent.

### *Critical Micelle Concentration*

Determine how much surfactant is required to prepare one litre of solution which meets the critical micelle point for this surfactant.

$$\frac{\text{PPM} \times \text{density}\left(\frac{\text{kg}}{\text{m}^3}\right)}{1000} = \text{mg/L}$$

First, get density into the right unit. This will mostly involve converting g/mL into kg/m³

Long form unit conversion

Density = 1.14 g/mL

$$\frac{1.14\,\cancel{\text{g}}}{\cancel{\text{mL}}} \times \frac{1\,\text{kg}}{1000\,\cancel{\text{g}}} \times \frac{1000\,\cancel{\text{mL}}}{1\,\cancel{\text{L}}} \times \frac{1\,\cancel{\text{L}}}{0.001\,\text{m}^3} = 1140\,\text{kg/m}^3$$

What you'll notice is that the conversions from grams to kilograms and millilitres to litres cancel out. This allows for the shortcut:

$$114\,\text{g/mL} \times 1000 = 1140\,\text{kg/m}^3$$

CMC Dow Ecosurf EH-3 is given as 480 ppm

Density $0.9651\,\text{g/mL} = 965.1\,\text{kg/m}^3$

$$\frac{\text{PPM} \times \text{density}\left(\frac{\text{kg}}{\text{m}^3}\right)}{1000} = \text{mg/L}$$

$$\frac{480 \times 965.1\,\text{kg}/\text{m}^3}{1000} = 463.248\,\text{mg}/\text{L}$$

$$463.248\,\text{mg}/\text{L} = 0.463\,\text{g}/\text{L}$$

Approximately 0.5 g of the current formulation Dow EcoSurf EH-3 per litre of water will reach the critical micelle point for this surfactant. As the literature suggests, that efficient detergency is found in the range of 3–5 times a given CMC, the true starting range corresponds to 1.4 to 2.4 g/L for the current formulation.

***pH Buffer Quantity***

*Equations and Terms*

Henderson–Hasselbalch equation

$$pH = pKa + log_{10}\left(\frac{[A-]}{[HA]}\right)$$

Reorganised without logarithms:

$$K_a = \frac{[A-][H+]}{[HA]}$$

$K_a$ = The acid dissociation constant
[A−] = The concentration of conjugate base
[H+] = concentration of conjugate acid
[HA] = concentration of an acid

*Maths: Example*

Determine how much acetic acid is required to make a solution of pH 4.2

$$K_a = \frac{[A-][H+]}{[HA]}$$

1. Look up the acid dissociation constant for the given acid. This is usually available in chemistry reference literature and websites such as PubChem. $K_a$ of acetic acid $1.76 \times 10^{-5}$
2. Calculate the ratio between the conjugate acid/base pair, here focussing on the proton [H+] using the desired pH of the end solution.

$$\text{pH} = -\log([\text{H}+])$$

$$[\mathrm{H}+]=10^{-\mathrm{pH}}$$

Desired pH is 4.2

$$[\mathrm{H}+]=10^{-4.2}$$
$$[\mathrm{H}+]=6.3\times10^{-5}$$

3. Insert the researched $K_a$ value for acetic acid and the acid concentration [H+] calculated from the desired pH into the reorganised Henderson–Hasselbalch equation.

$$K_a=\frac{[A-][\mathrm{H}+]}{[\mathrm{HA}]}$$

$$1.76\times10^{-5}=\frac{[\mathrm{CH3COO}]6.3\times10^{-5}}{[\mathrm{CH3COOH}]}$$

$$\frac{1.76\times10^{-5}}{6.3\times10^{-5}}=\frac{[\mathrm{CH3COO}-]}{[\mathrm{CH3COOH}]}$$

$$0.28=\frac{[\mathrm{CH3COO}-]}{[\mathrm{CH3COOH}]}$$

0.28 = molar concentration of acetate to acetic acid (mol/L).

So for each 100 mL of 1 M acetic acid you will need 28 mL of 1 M acetate to reach a solution of pH 4.2.

#### *Enzyme Activity*

Equations adapted from Erickson 1992 [23]

##### *Equations and Terms*

$$\mathrm{m_s}=\frac{\mathrm{C_V}\times\mathrm{V}}{\mathrm{A_s}}$$

$m_s$ = the mass of solid enzyme-containing materials desired
$C_V$ = the concentration (units/mL) desired activity
V = volume of the solution
$A_s$ = the units of activity per milligram of solid. Note the distinction between solid and enzyme.

**Example**

Prepare a 100 mL solution at the concentration of 25 u/mL *Aspergillus oryzae* with and activity of 30 u/mg

$$m_s = \frac{\left(\frac{25\ u}{mL} \times 100\ mL\right)}{\frac{30u}{mg}}$$

$$m_s = 83\ mg$$

$$m_s = 83\ mg \times \frac{1\ g}{1000\ mg}$$

$$m_s = 0.083\ g$$

## Health and Safety

Consultant: Barry Knight

For those who carefully read the previous section, please now disregard everything stated about the relativity of 'best' practice. Although specific handling and disposal methods will vary based on local legislation, when it comes to health and safety, this is the one section where no compromises should be made, no corners cut.

Unfortunately, the ever-evolving nature of health and safety research and legislation means that it is impossible at best (or dangerously misleading at worst) to write comprehensively on the topic. Instead, the following section will aim to outline a few key ideas around protective equipment and to point the reader towards reputable resources. This section is very much geared towards UK and European standards but will hopefully provide information that is useful in many other locations.

### *Solvents*

A very quick note about solvents. With increased environmental awareness and the outright banning of many once commonplace solvents, some such materials are swiftly (and rightly) disappearing from conservation laboratories [24].

Although the long-term effects of solvent exposure generally are a subject of evolving research, there have also been considerable gains in the development and characterisation of 'green solvents'. The term 'green solvents' is widely used in chemical and industrial literature, but given the popular use of 'green' ad absurdum, this text will adopt Fife and Gill-Brown's convention of 'greener' [25]. This focus on saf*er* and bett*er* gets at the relativity of these ideas and feels like a more honest appraisal of the materials than safe, best, or green.

Green(er) solvent research looks not only at the physiological effects of exposure, but also takes into account broader environmental effects related to use, production, and disposal, as well as general safety considerations (such as flammability). A particularly helpful report has been produced by Innovative Medicines Initiative (IMI) CHEM21, a European consortium of pharmaceutical companies, universities, and other related enterprises [26]. The resulting guidance is freely available and easy to interpret, with clearly adaptable methodologies. Thankfully, due to this last point, although the CHEM21 Green Solvents report does not specifically address some materials commonly used in conservation treatments, the transparency of their methods allows them to be applied. Table 1.2 collates information from the CHEM21 report and applies the methodology to several additional solvents.

To distinguish solvents unique to this text from those directly discussed in the report, Table 1.2 includes terminology from the report as well as a 'concern level' devised for this publication. CHEM21 terms used in the report: R – Recommended, P – Problematic, H – Hazardous, HH – Highly Hazardous, whereas this publication opts for the even less specific 'levels of concern' from least (1), to most (4) [26].

A notable addition to this table is white spirit, as the composition of this solvent is infamously ambiguous. What species of hydrocarbons are present is poorly defined, regionally variable, and completely dependent on the source, but percent composition of aromatic components can still range anywhere from 2–28% in just one source [27–29]. As ever, the SDS for each solvent should be referenced prior to use, but additional care with non-specific white spirit, mineral spirits, naphtha, and petroleum spirits is suggested. Although a bit out of date, Alan Phenix's excellent WAAC article carefully unpicks the nomenclature around all the petroleum-derived 'spirits' and is a worthwhile read [30].

If this publication is fortunate enough to see future revision, I sincerely hope that this section will be thickly populated with replacement solvents and solvent blends that are only just now being explored. The Greener Solvents Project run by Sustainability in Conservation is particularly notable, and readers are strongly encouraged to see both their current handbook and publications [31].

### *Personal Protective Equipment (PPE)*

For those unfamiliar with the abbreviation, personal protective equipment (PPE) includes all manner of safety wear; from hard hats and high visibility vests, to face shields, lab coats, and respirators [32]. PPE is typically carefully controlled by local legislative authorities, and as a result the labelling conventions, terminology, and symbols can differ from nation to nation and region to region. The following text will focus on conventions within the EU and UK, but will indicate towards similar regulatory bodies in some other countries where possible.

*Table 1.2* Summary of some solvents discussed in the CHEM21 report, including those historically used in conservation practice. See the CHEM21 report for definitions on the scoring [26]

| *Class* | *Solvent* | *CAS* | *Safety Score* | *Health Score* | *Environmental Score* | *CHEM 21 Rating* | *Concern 1 least to 4 most* |
|---|---|---|---|---|---|---|---|
| Alcohol | Benzyl Alcohol | 100-51-6 | 1 | 2 | 7 | Problematic | 2 |
| Alcohol | Butanol | 71-36-3 | 3 | 4 | 3 | Recommended | 1 |
| Alcohol | Ethanol | 64-17-5 | 4 | 3 | 3 | Recommended | 1 |
| Alcohol | methanol | 67-56-1 | 4 | 7 | 5 | Recommended | 2 |
| Alcohol | propanol | 67-63-0 | 4 | 3 | 3 | Recommended | 1 |
| Amine | Ammonia | 7664-41-7 | 1 | 7 | 7 | - | 3 |
| Amine | Pyridine | 110-86-1 | 4 | 2 | 3 | Hazardous | 3 |
| Amine | Triethanolamine (TEA) | 102-71-6 | 6 | 7 | 3 | Hazardous | 3 |
| Aprotic polar | 1-methyl-2-pyrrolidinone | 872-50-4 | 1 | 9 | 7 | Hazardous | 3 |
| Chlorinated | Carbon Tetrachloride | 56-23-5 | 2 | 7 | 10 | **Highly Hazardous** | 4 |
| Chlorinated | Chlorobenzene | 108-90-7 | 3 | 2 | 7 | Problematic | 2 |
| Chlorinated | Chloroform | 67-66-3 | 2 | 7 | 5 | **Highly Hazardous** | 4 |
| Chlorinated | Dichloromethane (DCM) | 75-09-2 | 1 | 7 | 7 | Hazardous | 3 |
| Ester | Ethyl Acetate | 141-78-6 | 5 | 3 | 3 | Recommended | 1 |
| Ether | Diethyl Ether | 60-29-7 | 10 | 3 | 7 | **Highly Hazardous** | 4 |
| Ether | Tetrahydrofuran (THF) | 109-99-9 | 6 | 7 | 5 | Problematic | 2 |
| Ether | 1-methoxy-2-propanol | 107-98-2 | 3 | 2 | 1 | - | 1 |
| Hydrocarbons | Benzene | 71-43-2 | 6 | 10 | 3 | **Highly Hazardous** | 4 |
| Hydrocarbons | Cyclohexane | 110-82-7 | 6 | 3 | 7 | Problematic | 2 |
| Hydrocarbons | Heptanes | 142-82-5 | 6 | 2 | 7 | Problematic | 2 |
| Hydrocarbons | Hexanes | 110-54-3 | 8 | 7 | 7 | Hazardous | 3 |
| Hydrocarbons | Naphtha | 8030-30-6 | 7 | 2 | 7 | - | 3 |
| Hydrocarbons | Pentane | 109-66-0 | 8 | 3 | 7 | Hazardous | 3 |
| Hydrocarbons | petroleum Spirits | - | 4 | 2 | 7 | - | 2 |
| Hydrocarbons | Shellsol D38 | 64742-48-9 | 3 | 3 | 7 | - | 2 |
| Hydrocarbons | Turpentine | 8006-64-2 | 3 | 3 | 7 | - | 2 |
| Hydrocarbons | white spirit | | 3 | 6 | 7 | - | 3 |
| Hydrocarbons | Xylene | 1330-20-7 | 4 | 2 | 5 | Problematic | 2 |
| Ketone | Acetone | 67-64-1 | 5 | 3 | 5 | Recommended | 1 |
| Ketone | Methyl Ethyl Ketone (MEK) | 78-93-3 | 5 | 3 | 3 | Recommended | 1 |

At this point it's worth reiterating the importance of ordering through legitimate suppliers. Despite the severity of penalties for suppliers if caught, the availability of counterfeit PPE has risen significantly in the years immediately following the COVID-19 pandemic [2]. It's therefore worth belabouring the earlier point (*Overview*) about the importance of sourcing PPE from specialist retailers and not simply going to the cheapest online retailer.

*Labelling: Standard Markings*

In most regions, legitimate PPE will be clearly marked. This is to designate which materials meet the established 'conformity' levels of the relevant verifying body. At the time of writing, the British government retains the same standards for PPE as the rest of Europe, with official guidance stating that the government 'intends to legislate to continue recognition of EU requirements … indefinitely for a range of product regulations' [33].

PPE sold within the EU and UK should have a 'CE' emblem (Figure 1.2) and an EN number printed clearly on the packaging and/or equipment. The Conformité Européenne (CE) mark indicates that the equipment meets European standards generally and the Europäische Norm (EN) number gives the reference for the relevant statute. Although the UK is in agreement with European standards for PPE, Brexit has seen the introduction of an additional UK-specific conformity marking or UKCA logo (Figure 1.3) [33].

*Eye Protection*

The rating and protective ability of a given piece of eyewear will typically be inscribed on the inside of one arm and on the lens itself. These markings include information about production, filtration, and other characteristics and explanations will be available through supplier websites. As with any PPE, the

CE

*Figure 1.2* CE mark from the European Commission [34].

UK
CA

*Figure 1.3* UKCA mark from UK department for business and trade [33].

right equipment should be selected for the job at hand. In Europe, protective eyewear must meet the standards set out in EN166. This statue meets requirements for both impact and splash protections and therefore covers everything from welders' masks to full-face guards [32].

In terms of fit, the Canadian CCOHS provides excellent general advice, stating that eyewear should be individually assessed to minimize any gaps between the sides of the device and the wearer's face. Ideally, protective eyewear should cover the entire orbital region, from the top of the cheekbone up and over the eyebrow and fully across the bridge of the nose. Finally, it's important that any eyewear be well fitting and comfortable, suitable for the specific face size and shape of the user, and able to be worn over any prescription glasses. Glasses that slide down your nose won't protect your eyes, and goggles that aren't worn because they cut into your face will be no help at all [35].

### *Gloves*

This section was originally intended to tabulate clear guidance for what kinds of gloves are best suited for work with various common solvents. Unfortunately, however, as is often the case with PPE, there is no simple such advice to be given. However, I hope that summarising the key issues will provide necessary context for navigating the complexity of materials and labelling.

#### KEY IDEA

If you are working with a hazardous material, you must pick the glove that is specifically tested to offer protection for that precise material. Do not assume that the thin nitrile gloves that are so useful in object handling offer *any protection at all* against chemicals and solvents.

Conservation spaces equipped with solvents other than water should strongly consider purchasing different glove types for different materials.

#### GLOVE FAILURE IN THEORY

As anyone who has attempted to use nitrile examination gloves with acetone can tell you, not all gloves are equally protective for all materials. This variation stems from both the thickness of the glove as well as the exact material (or combination of materials) it is made from. A short summary of how failure is understood and described is therefore required before discussion of mitigation can be begun. There are three key ideas for this: degradation, breakthrough time, and permeation rate.

Degradation, as you may expect, refers to the physical changes to the material and can take the form of shrinking, swelling, cracking, or shredding (among others) [36]. Breakthrough time and permeation rate are perhaps more helpful concepts as these refer to how quickly materials can penetrate a glove. Breakthrough time describes how long from the point of exposure that it takes for the material outside the glove to be detected inside the glove. Permeation rate is the speed at which a substance passes through a glove once breakthrough has occurred. Please note that permeation and breakthrough can occur without clear indications of degradation [36].

### GLOVE FAILURE IN PRACTICE

This is where things can be a little concerning. Despite the severity of potential consequences of mishandling hazardous materials, the industrial literature is dispiriting. According to several studies published in the *Journal of Occupational and Environmental Hygiene* and the *Journal of Applied Polymer Science*, the "variability in chemical permeation among brands is high, with up to 10-fold differences in BT [breakthrough] and 200-fold differences in PR [permeation rate]" [37–39]. This means that the same types of gloves, made with the same materials and tested to the same standards exhibited enormous variation both in the initial time to break through and the rate the material passed into the glove once breakthrough had occurred.

Furthermore, not only does the protection vary from brand to brand, Phalen, Le and Wong also found that the movement of gloved hands can also diminish their protective ability [40]. For example, a nitrile glove which offers good chemical resistance to ethanol had the breakthrough time drop by nearly one-third and its overall permeation rate raised by almost half (47%) when exposed to simple repeated hand movements [40].

This has led at least one researcher and industrial hygienist to conclude that general guidelines may not reflect true performance. In the absence of product testing, "assume the level of protection may be less than expected" [37].

Now, conservators are (hopefully) rarely encountering highly hazardous materials, and direct physical contact with any chemicals or solvents should be largely avoided through the use of various hand tools (pipettes, swabs, spatula, etc.). This information is included to therefore urge caution and to provide context for why several types of gloves might be a worthwhile addition to a conservation space. We know that gloves don't always function as well as intended for the materials they are tested against, so they certainly can't be relied upon to protect the user from material they are not designed for.

How then to make sensible choices? Now that the problems are clear, the following sections collate information intended to inform specific decision-making as well as provide context for interpreting some conventions in labelling.

### MATERIALS

The first thing to consider is materials the glove is made from as some polymers or polymer blends will exhibit inherent resistance to some solvents over others. The information in Table 1.3 is compiled from recent guidance prepared by the American Occupational Safety and Health Administration (OSHA), roughly the equivalent of the British Health and Safety Executive (HSE). Please note

*Table 1.3* Chemical resistance of glove materials to various industrial chemicals, abridged from full report available online [41]

| *Chemical* | *Neoprene* | *Latex/Rubber* | *Butyl* | *Nitrile* |
|---|---|---|---|---|
| Acetaldehyde | Very Good | Good | Very Good | Good |
| Acetic acid | Very Good | Very Good | Very Good | Very Good |
| Acetone | Good | Very Good | Very Good | Poor |
| Benzene | Poor | Poor | Poor | Fair |
| Hexane | Fair | Poor | Poor | Good |
| Hydrochloric acid | Very Good | Good | Good | Good |
| Hydrogen peroxide (30%) | Good | Good | Good | Good |
| Isooctane | Fair | Poor | Poor | Very Good |
| Ketones | Good | Very Good | Very Good | Poor |
| Methanol | Very Good | Very Good | Very Good | Very Good |
| Methyl ethyl ketone | Good | Good | Very Good | Poor |
| Napthas, aliphatic | Very Good | Fair | Fair | Very Good |
| Napthas, aromatic | Good | Poor | Poor | Good |
| Isopropyl alcohol | Very Good | Very Good | Very Good | Very Good |
| Tannic acid (65) | Very Good | Very Good | Very Good | Very Good |
| Tetrahydrofuran | Poor | Fair | Fair | Fair |
| Toluene* | Fair | Poor | Poor | Fair |
| Triethanolamine (85%) | Very Good | Good | Good | Very Good |
| Turpentine | Good | Fair | Fair | Very Good |
| Xylene* | Poor | Poor | Poor | Fair |

Note that many solvents used in conservation have not been specifically tested in these reports. Chemicals marked with an asterisk (*) are for "limited service".

that this provides general guidance and does not take into account the effects of thickness or the variability described in the previous section.

#### THICKNESS

While thicker gloves generally offer better chemical protection than thinner, this is not linear nor is it significant when the changes are small [37]. While doubling or tripling the thickness of a glove can cause significant slowing in deterioration and breakthrough times, changing from 5 mil to 6 mil is unlikely to have any noticeable effect [37].

Confusingly, please note that despite its pronunciation, a 'mil' is not the same as millimetre (mm), but is instead a colloquial imperial-metric hybrid word that essentially works as a milli-inch, or 1/1000th of an inch [42].

#### MARKINGS AND STANDARDS

As discussed previously, labelling legislation varies regionally, and conservators working outside the EU and UK are encouraged to seek guidance from the relevant legislative body. In Europe, gloves designed to protective standards must be labelled with the type and the chemicals they were tested against. This labelling system is described by the standard EN ISO 374-1:2016 and gives

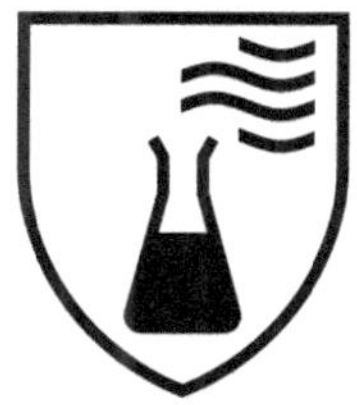


*Figure 1.4* Annotated version of ISO 7000-2414 International Organization for Standardization (ISO) [43].

*Table 1.4* Definition of 'type' for EN 16523-1:2015 [45]

| *Type* | *Number of chemicals* | *Breakthrough time* |
|---|---|---|
| A | 6 or more | At least 30 minutes |
| B | 3 to 5 | At least 30 minutes |
| C | Less than 3 | 10–29 minutes |

explicit information about the ability of a specific glove type to protect against a range of solvents over a discrete amount of time (Figure 1.4).

In this image, the statue number and 'type' are listed above the central pictogram of a steaming beaker, and there are a series of letters underneath. The capital letters below the beaker refer to specific chemicals, and the "Type" describes the breakthrough rates for the same. Please note that although some information can be extrapolated from these figures, the chemicals listed are the only ones tested, and resistance to anything else unknown and therefore unadvised (Tables 1.4 and 1.5) [44, 45].

SUMMARY

- Designate gloves for touch protection, and those for chemical protection. Ensure others are aware of the distinction.
- Keep a few different types of gloves and label them clearly. Neoprene in particular shows better resistance to alcohols and ketones than nitrile. Suppliers' websites typically make this explicit, with solvent search functions by name and CAS number.
- Readily switch between touch protection and chemical protection as needed.
- Although you may be wearing gloves, act as if you aren't. Avoid direct contact with solvents (for example, using a rolled swab or brush rather than cotton wool held between fingers).
- Assume that thin nitrile offers very little chemical protection.
- When working with solvents, replace gloves regularly and do not attempt to reuse.

*Table 1.5* The representative letters for each chemical used for permeation testing according to EN 16523-1:2015 [45]

| | *Chemical* | *CAS number* | *Type* |
|---|---|---|---|
| A | Methanol | 67-56-1 | Primary alcohol |
| B | Acetone | 67-64-1 | Ketone |
| C | Acetonitrile | 75-05-8 | Nitrile Compound |
| D | Dichloromethane | 75-09-2 | Chlorinated hydrocarbon |
| E | Carbon disulphide | 75-15-0 | Sulphur containing organic compound |
| F | Toluene | 108-88-3 | Aromatic hydrocarbon |
| G | Diethylamine | 109-89-7 | Amine |
| H | Tetrahydrofuran | 109-99-9 | Heterocyclic and ether compound |
| I | Ethyl acetate | 141-78-6 | Ester |
| J | n-Heptane | 142-82-5 | Saturated hydrocarbon |
| K | Sodium hydroxide 40% | 1310-73-2 | Inorganic base |
| L | Sulphuric acid 96% | 7664-93-9 | Inorganic mineral acid, oxidizing |
| M | Nitric acid 65% | 7697-37-2 | Inorganic mineral acid, oxidizing |
| N | Acetic acid 99% | 64-19-7 | Organic acid |
| O | Ammonia 25% | 1336-21-6 | Organic base |
| P | Hydrogen peroxide 30% | 7722-84-1 | Peroxide |
| S | Hydrofluoric acid 40% | 7664-39-3 | Inorganic mineral acid |
| T | Formaldehyde 37% | 50-00-0 | Aldehyde |

Many of the listed materials are not used in conservation practice.

### *Masks/Respirators*

As with the section on gloves, this section cannot corral the entire legislative swarm that surrounds all PPE into simple, pithy statements. Instead, a few key ideas will be discussed, and particularly helpful references will be referenced.

#### KEY IDEAS

If you will forgive the repetition, note that when working with a hazardous material, you must pick the respiratory protection that is specifically designed for that precise material and has been tested for the individual using it. For example, if working in the presence of mercury, shockingly large quantities of which can be found in historic barometers, you must use a respirator that is specifically designed to filter mercury vapour.

Filters have limited shelf life after which they will not function as advertised. Check expiry and longevity dates and dispose of out-of-date materials, even if completely unused.

As each individual face is different, one cannot expect that every mask will seal well against every face. This can only be determined by individual testing, known as face fit testing. In the UK, face fit testing is a legal requirement, to be determined by a specialist 'competent person' and cannot consist of a simple seal check [46].

Please note that a half-mask will simply not form a close seal on a face with whiskers. Those with beards or other facial hair will be required to shave to effectively use a mask.

From personal experience, a well-fitting mask feels tight, even mildly uncomfortable when worn for long periods and will almost certainly leave temporary welts on your face.

Masks must be sourced from reputable suppliers and checked for correct labelling. Although certainly less of an issue now, the COVID-19 pandemic saw an unfortunate influx of counterfeit masks. In 2020 the Emergency Care Research Institute (ECRI) found that 60–70% of tested masks which were marketed as KN95 did not meet minimum standards for the type [47, 48]. The American Centers for Disease Control (CDC) website is full of good information on the topic, but now that supplies are less stretched, it remains easier to simply buy protective equipment from a reputable supplier.

### MARKINGS AND STANDARDS

In Europe, the relevant standards for half-mask particulate respirators are EN140 and EN149, and one of these numbers will be clearly marked on RPE packaging. See Table 1.4 for a collection of similar RPE standards and regulatory bodies.

Importantly for particulate work, the conformity standard EN149 can then be further subdivided based on the level of filtration the mask can perform. For dust masks, this is listed as a Filtering Face Piece (FFP) number, with higher numbers corresponding to greater filtration. For example: FFP1, FFP2, and FFP3 respirators can reduce the amount of particulates by a factor of 4, 10, and 20 respectively [49].

In the US RPE is divided first by oil resistance; N for no resistance, R for resistant up to 8 hours, and P is completely oil-resistant. This is then followed by a numerical value describing the percentage of particles filtered [50]. An N95 will therefore offer no resistance to oil, but will filter 95% of particles.

Manufacturers websites are full of information about which respirators are best for which hazards in what quantities. When looking to purchase RPE, research your needs directly through the manufacturer and ensure, again, that you purchase from reputable sources. In November 2021, the *New York Times* cited a glut of counterfeit KN95s available on Amazon [51]. The company has more recently met with European Commissioner for the Internal Market Thierry Breton to outline their new approach to combat the availability of counterfeit materials on their marketplace, but following the waning of the pandemic, no further corroborative research into the topic has been forthcoming (Table 1.6) [52].

*Table 1.6* Standards and regulatory bodies for respiratory protective equipment in selected countries

| *Nations or Region* | *Standard* | *Important Regulatory terms or bodies* |
|---|---|---|
| Australia/New Zealand | AS/NZA 1716:2012 | Safe Work Australia<br>WorkSafe New Zealand |
| Brazil | ABNT/NBR 13.698-2011 | Associação Brasileira de Normas Técnicas (ABNT)<br>Norma Brasileira (NBR) |
| Canada | CSA Standard Z94.4-11 (R2016) | CSA Certification<br>Canadian Centre for Occupational Health and Safety (CCOHS) |
| China | GB2626-2006 | Guobiao (GB)<br>The Standardization Administration of China |
| Europe | EN 149-2001 | Europäische Norm/Euronorm (EN)<br>European Committee for Standardization |
| Japan | JMHLW-Notification 214, 2018 | Japan Ministry of Health, Labour, and Welfare (JMHLW) |
| Korea | KMOEL – 2017-64 | KC Safety Certification<br>Korean Agency for Technology and Standards (KATS) |
| United States | NIOSH-42CFR84 | The National Institute for Occupational Safety and Health (NIOSH)<br>Occupational Safety and Health Administration (OSHA) |

***General Reference Materials***

*Table 1.7* Reference materials. Rather than provide a table of unclickable urls and links that would no doubt break within a year, I've instead listed the names and a short description of important resources

| *Topic* | *Name* | *Short description* |
|---|---|---|
| Analytics | Equipment Data JISC | An analytical equipment database showing which techniques (SEM, GCMS, etc.) are available at which UK research organisations and universities. Operated by the UK's Joint Information Systems Committee (JISC) |
| Analytics | Infrared & Raman Users Group (IRUG) | Heritage-focussed searchable database of comparative reference spectral data from Raman and infrared spectroscopy |
| Calculations | Merck/Sigma-Aldrich Buffer Reference Center | Helpful technical information for buffer types, pH ranges and preparation tables |

(*Continued*)

*Table 1.7* (Continued)

| *Topic* | *Name* | *Short description* |
|---|---|---|
| Calculations | Molarity calculator | There are several of these, I've found those run by Sigma-Aldrich and OmniCalculator to be the most intuitive. |
| Calculations | Omni Calculator | Website dedicated to calculations grouped by discipline or theme such as 'chemistry' or 'construction' |
| Chemical Info | IUPAC Compendium of Chemical Terminology "Gold Book". | One of several reference works prepared by International Union of Pure and Applied Chemistry (IUPAC). The Gold Book contains internationally accepted definitions for chemical terms. Currently available as a website |
| Chemical Info | National Institute of Standards and Technology (NIST) Chemistry WebBook | Searchable website iteration of the NIST Standard Reference Database Number 69. Includes general chemical information, including structures and links to other thermochemical data and spectra from various spectroscopy methods. |
| Chemical Info | PubChem | PubChem is a publicly available website database of chemicals containing information about various properties and biological assay activity. This database is operated by the US National Center for Biotechnology Information (NCBI) which is part of the National Institutes of Health (NIH). |
| Chemical Info | ChemSpider | A publicly available database of chemical information operated by the UK Royal Society of Chemistry. |
| Chemical Safety | NIOSH Pocket Guide to Chemical Hazards | Website devoted to providing chemical and safety information to US standards. Includes specific respirator recommendations. |
| Chemical Safety | International Labour Organization: International Chemical Safety Cards (ICSCs) | United Nations agency devoted to the promotion of human and labour rights. The 'cards' list chemical safety information including clear hazard information and safety measures. |
| Conservation Reference | Conservation & Art Materials Encyclopedia Online (CAMEO) | Database and reference library of conservation related materials and terms. |
| Conservation Reference | Art and Archaeology Technical Abstracts (The Getty's AATA) | Database collecting abstracts from conservation-related publications. |

(*Continued*)

*Table 1.7* (Continued)

| *Topic* | *Name* | *Short description* |
|---|---|---|
| Conservation Reference | 'AIC's Book and Paper Group Recipes | A community run wiki from the American Institute of Conservation's Book and Paper group. This includes recipes and tips for many techniques and materials used in book/paper. |
| Conservation Reference | Conservation Online (CoOL) | Enormous collection of articles, threads, and newsletters containing a wealth of information related to conservation. Run by the Foundation for Advancement in Conservation (FAIC) |
| Conservation Reference | Plastic Identification Tool | Context-based tool for the identification of plastic objects, containing information on degradation pathways and storage recommendations grouped by polymer type. Developed by the Dutch Cultural Heritage Agency Rijksdienst voor het Cultureel Erfgoed. |
| Conservation Reference | Canadian Conservation Institute (CCI) Notes | Series of short reports detailing specific challenges and recommendations in the care of cultural heritage. Detailed, practical, and well-sourced. |
| Conservation Reference | Museums Galleries Scotland – Collections advice | Aimed at small museums but contains good summary of general standards and UK-specific topics such as Government Indemnity Scheme (GIS) requirements. |
| General H&S | Health and Safety Executive (HSE) | UK governmental body tasked with regulating and enforcing workplace welfare. The hse.gov.uk website has a 'guidance' section which has information on a broad range of topics including PPE, working at height, and many others. This body exists in many countries, and local rules apply |
| General H&S | Hazards in Collections | Digital tool designed to help identify hazardous materials commonly found in museum collections. Contains many photos of examples and information about likely risks. |

*Table 1.8* Temperature conversions

| *C* | *-20* | *-15* | *-10* | *-5* | *0* | *5* | *10* | *15* | *20* | *25* | *30* | *35* | *40* | *45* | *50* | *55* | *60* | *65* | *70* | *75* | *80* | *85* | *90* | *95* | *100* |
|---|---|---|---|---|---|---|---|---|---|---|---|---|---|---|---|---|---|---|---|---|---|---|---|---|---|
| F | -4 | 5 | 14 | 23 | 32 | 41 | 50 | 59 | 68 | 77 | 86 | 95 | 104 | 113 | 122 | 131 | 140 | 149 | 158 | 167 | 176 | 185 | 194 | 203 | 212 |
| F | -20 | -10 | 0 | 10 | 20 | 30 | 40 | 50 | 60 | 70 | 80 | 90 | 100 | 110 | 120 | 130 | 140 | 150 | 160 | 170 | 180 | 190 | 200 | 210 | 220 |
| C | -29 | -23 | -18 | -12 | -7 | -1 | 4 | 10 | 16 | 21 | 27 | 32 | 38 | 43 | 49 | 54 | 60 | 66 | 71 | 77 | 82 | 88 | 93 | 99 | 104 |

*Table 1.9* Dilution (V:V)

| | *50%* | *20%* | *10%* | *5%* | *4%* | *2%* | *1%* |
|---|---|---|---|---|---|---|---|
| Ratio (v:v) | 1:1 | 1:4 | 1:9 | 1:19 | 1:24 | 1:49 | 1:99 |
| Solute | 50 | 20 | 10 | 5 | 4 | 2 | 1 |
| Solvent | 50 | 80 | 90 | 95 | 96 | 98 | 99 |
| Total | 100 | 100 | 100 | 100 | 100 | 100 | 100 |

*Table 1.10* UK and US imperial to metric volume conversion [53, 54]

| | | | | | |
|---|---|---|---|---|---|
| Imperial (UK) | 1 tsp | 1 Tbsp | 1 fl oz | 1 cp | 1 pint |
| metric | 5 mL | 15 mL | 30 mL | 240 mL | 500 mL |
| Imperial (US) | 1 tsp | 1 Tbsp | 1 fl oz | 1 cp | 1 pint |
| metric | 4.93 mL | 14.79 | 29.57 mL | 240 mL | 470 mL |

## References


[1] J.H. Townsend, *An Introduction To Materials*, 3rd ed. Science for conservators series, Volume 1. Milton Park, Abingdon, Oxon; New York, NY: Routledge, 2023.

[2] OECD and European Union Intellectual Property Office, *Dangerous Fakes: Trade in Counterfeit Goods that Pose Health, Safety and Environmental Risks.* in Illicit Trade. OECD, 2022. doi: 10.1787/117e352b-en

[3] Know The Chain, "The issue," Know the Chain [Online]. Available: https://knowthechain.org/the-issue/

[4] D. Pinkert and C. Tate, "Stringency of US forced labor law is crucial to its success," *Sourcing Journal*, Nov. 27, 2023. [Online]. Available: https://sourcingjournal.com/topics/thought-leadership/us-forced-labor-law-china-uflpa-customs-cbp-uyghur-genocide-469337/

[5] European Commission, "Just and sustainable economy: Commission lays down rules for companies to respect human rights and environment in global value chains." Feb. 23, 2022. [Online]. Available: https://ec.europa.eu/commission/presscorner/detail/en/IP_22_1145

[6] UK Home Office, "Transparency in supply chains etc. A practical guide: Guidance issued under section 54(9) of the Modern Slavery Act 2015," Government of the United Kingdom, Guide, 2015. [Online]. Available: https://assets.publishing.service.gov.uk/media/61b7401d8fa8f5037778c389/Transparency_in_Supply_Chains_A_Practical_Guide_2017_final.pdf

[7] T. Higgins, "Publication Contribution - Conservator's Cookbook," [Personal Communication] May 29, 2024.

[8] C. Landström, R. Hauxwell-Baldwin, I. Lorenzoni, and T. Rogers-Hayden, "The (Mis)understanding of Scientific Uncertainty? How Experts View Policy-Makers, the Media and Publics," *Science as Culture*, vol. 24, no. 3, pp. 276–298, 2015, doi: 10.1080/09505431.2014.992333

[9] M. Hicks and G. Harris, "Specifications and tolerances for reference standard and field standard weights and measures," National Institute of Standards and Technology, NIST Handbook 105-2, 2022. doi: 10.6028/NIST.HB.105-2e2021

[10] ASTM International, "Standard specification for laboratory glass graduated cylinders." 2019. doi: 10.1520/E1272-02R19

[11] United Kingdom Accreditation Service (UKAS), "Traceability: Volumetric apparatus." Dec. 05, 2023. [Online]. Available: https://www.ukas.com/wp-content/uploads/2023/05/LAB-15-Traceability-Volumetric-apparatus.pdf

[12] "Transfer Pipettes, FINN Pipette Tips, in various sizes," Vetlab Supplies Ltd. Accessed: Apr. 13, 2024. [Online]. Available: https://vetlabsupplies.co.uk/products/laboratory-consumables/transfer-pipettes-pipette-tips/

[13] Fisher Scientific, "Transfer Pipets," Fisher Scientific. Accessed: Apr. 13, 2024. [Online]. Available: https://www.fishersci.co.uk/gb/en/browse/90140055/transfer-pipets?page=1

[14] B. Krystal, "You can't always trust measuring cups, but here's what you can do about it," *Washington Post*, Oct. 28, 2021. Accessed: Mar. 23, 2024. [Online]. Available: https://www.washingtonpost.com/news/voraciously/wp/2019/03/04/you-cant-always-trust-measuring-cups-but-heres-what-you-can-do-about-it/

[15] Lab Compare, "Calibration best practices: precision balances and analytical balances," Lab Compare. Accessed: Mar. 24, 2024. [Online]. Available: https://www.labcompare.com/562607-Calibration-Best-Practices-Precision-Balances-and-Analytical-Balances/

[16] A. O'Driscoll, "A guide to balance and scale calibration," Lab Balances. [Online]. Available: https://labbalances.net/blogs/blog/balance-and-scale-calibration-guide

[17] J. Haller, E. Hage, N. Ludolph, and A. Taube, "Minimum sample weights according to USP <41>, OIML R 76 and EURAMET cg-18," Sartorius, White paper, Aug. 2020. [Online]. Available: https://www.sartorius.com/resource/blob/580758/6e89cff42ec6efa2863140b467dc5463/minimum-weight-white-paper-en-sartorius-pdf-data.pdf

[18] Precisa, "Accurate weighing | selecting the right balance | top tips," Precisa. Accessed: Apr. 13, 2024. [Online]. Available: https://www.precisa.com/blog/toptips-to-ensure-accurate-weighing/

[19] United Kingdom Accreditation Service (UKAS), "Guidance on the calibration of weighing machines used in testing and calibration laboratories." Nov. 07, 2022. [Online]. Available: https://www.ukas.com/wp-content/uploads/schedule_uploads/759162/LAB-14-Guidance-on-the-calibration-of-weighing-machines.pdf

[20] V. Miller, *Recommended guide for determining and reporting uncertainties for balances and scales*, National Institute of Standards and Technology, Gaithersburg, MD, NIST IR 6919, 2002. doi: 10.6028/NIST.IR.6919

[21] C. Stavroudis, "Pemulen® revised: pHuck the pH Meter," *WAAC Newsletter*, vol. 34, no. 2, p. 19, 2012.

[22] N. A. Metheny, E. M. Gunn, C. S. Rubbelke, T. F. Quillen, U. R. Ezekiel, and K. L. Meert, "Effect of pH test-strip characteristics on accuracy of readings," *Crit Care Nurse*, vol. 37, no. 3, pp. 50–58, 2017, doi: 10.4037/ccn2017199

[23] H. M. Erickson, "Usage recommendations for alpha-amylases: Maximizing enzyme activity while minimizing enzyme-artifact binding residues," in *The Book and Paper Group Annual*, pp. 24–33, 1992.

[24] ECHA, "Substances restricted under REACH - Annex XVII." Accessed: Apr. 24, 2024. [Online]. Available: https://echa.europa.eu/substances-restricted-under-reach

[25] G. R. Fife and I. Gill-Brown, "Introduction," in *Greener Solvents in Conservation: an introductory guide*, G. R. Fife, Ed., London: Archetype Publications, 2021. [Online]. Available: https://siconserve.wpengine.com/wp-content/uploads/2021/12/Greener-Solvents-in-Conservation-ONLINE-VERSION-v5.pdf

[26] D. Prat *et al.*, "CHEM21 selection guide of classical- and less classical-solvents," *Green Chem.*, vol. 18, no. 1, pp. 288–296, 2016, doi: 10.1039/C5GC01008J

[27] Bartoline, "White Spirit," Safety Data Sheet, 2018. [Online]. Available: https://www.bartoline.co.uk/wp-content/uploads/2018/08/BARTOLINE-White.-Spirit-SDS4652.pdf

[28] ScrewFix, "No Nonsense White Spirit," Safety Data Sheet, 2023. [Online]. Available: https://ulmysds.com/Admin/ViewDocument.aspx?ID=2a16e9fa-ae48-48dc-935d-f2e83c66244c&primaryReportId=0

[29] VWR, "White Spirit," Safety Data Sheet, 2019. [Online]. Available: https://uk.vwr.com/assetsvc/asset/en_GB/id/7898659/contents/7898659.pdf

[30] A. Phenix, "Generic hydrocarbon solvents: a guide to nomenclature," *WAAC Newsletter*, vol. 29, no. 2, pp. 13–22, 2007.

[31] G. R. Fife, Ed., *Greener solvents in conservation: an introductory guide*. London: Archetype Publications, 2021. [Online]. Available: https://siconserve.wpengine.com/wp-content/uploads/2021/12/Greener-Solvents-in-Conservation-ONLINE-VERSION-v5.pdf

[32] Great Britain: Health and Safety Executive (HSE), *Personal protective equipment at work: regulations 1992 (as amended)*, 4th ed. S.l.: HSE BOOKS, 2022. [Online]. Available: https://books.hse.gov.uk/gempdf/L25.pdf

[33] UK Department for Business and Trade, "Using the UKCA marking," GOV.UK. Accessed: Apr. 24, 2024. [Online]. Available: https://www.gov.uk/guidance/using-the-ukca-marking

[34] European Commission, "CE marking - European Commission." Accessed: Jun. 04, 2024. [Online]. Available: https://single-market-economy.ec.europa.eu/single-market/ce-marking_en

[35] Government of Canada, "CCOHS: Eye and Face Protectors." Accessed: May 29, 2024. [Online]. Available: https://www.ccohs.ca/oshanswers/prevention/ppe/glasses.html

[36] University of Pennsylvania's Office of Environmental Health and Radiation Safety (EHRS), "Fact Sheet: disposable nitrile gloves in chemical labs." Accessed: Apr. 24, 2024. [Online]. Available: https://ehrs.upenn.edu/health-safety/lab-safety/chemical-hygiene-plan/fact-sheets/fact-sheet-disposable-nitrile-gloves

[37] R. N. Phalen, "Not all gloves are created equal," The Synergist. Accessed: Apr. 03, 2024.[Online].Available:https://synergist.aiha.org/201812-not-all-gloves-are-created-equal

[38] R. N. Phalen and W. K. Wong, "Polymer properties associated with chemical permeation performance of disposable nitrile rubber gloves," *Journal of Applied Polymer Science*, vol. 132, no. 6, p. app.41449, 2015, doi: 10.1002/app.41449

[39] T. D. Klingner and M. F. Boeniger, "A Critique of assumptions about selecting chemical-resistant gloves: a case for workplace evaluation of glove efficacy," *Applied Occupational and Environmental Hygiene*, vol. 17, no. 5, pp. 360–367, 2002, doi: 10.1080/10473220252864969

[40] R. N. Phalen, T. Le, and W. K. Wong, "Changes in chemical permeation of disposable latex, nitrile, and vinyl gloves exposed to simulated movement," *Journal of Occupational and Environmental Hygiene*, vol. 11, no. 11, pp. 716–721, 2014, doi: 10.1080/15459624.2014.908259

[41] Occupational Safety and Health Administration (OSHA), "Personal protective equipment," U.S. Department of Labor, OSHA 3151-02R 2023, 2023. [Online]. Available: https://www.osha.gov/sites/default/files/publications/osha3151.pdf

[42] Omni International Corp, "Millimeter to Mil to Gram: change in the glove industry," Omni International. Accessed: Apr. 24, 2024. [Online]. Available: https://omnigloves.com/2021/06/16/mil-to-gram/

[43] "ISO - International Organization for Standardization," ISO. Accessed: Jun. 08, 2024. [Online]. Available: https://www.iso.org/home.html

[44] UVex, "EN 374: Modified standard for chemical protective gloves," uvex xpert-blog. Accessed: Apr. 24, 2024 [Online]. Available: https://www.uvex-safety.com/blog/en-374-modified-standard-for-chemical-protective-gloves/

[45] Guide Gloves, "EN ISO 374-1:2016," Guide Gloves. Accessed: Apr. 24, 2024 [Online]. Available: https://guidegloves.com/en/knowledge/our-products/standards/en-iso-374-1-2016

[46] Health and Safety Executive (HSE), "Guidance on respiratory protective equipment (RPE) fit testing," INDG479, Mar. 2019 [Online]. Available: https://www.hse.gov.uk/pubns/indg479.pdf

[47] ECRI, "Up to 70% of Chinese KN95 masks tested by ECRI don't meet minimum standards," ECRI. Accessed: Apr. 24, 2024 [Online]. Available: https://d84vr99712pyz.cloudfront.net/p/images1/ecri-trusted-voice-healthcare.jpg

[48] NPPTL | NIOSH | CDC, "Counterfeit Respirators," The National Personal Protective Technology Laboratory (NPPTL). Accessed: Apr. 24, 2024 [Online]. Available: https://www.cdc.gov/niosh/npptl/usernotices/counterfeitResp.html

[49] UK Health and Safety Executive (HSE), "All you need to know about disposable dust masks," UK Health and Safety Executive (HSE). Accessed: Apr. 24, 2024 [Online]. Available: https://www.hse.gov.uk/agriculture/dustmasks.htm

[50] 3M, "Comparison of FFP2, KN95, and N95 filtering facepiece respirator classes," Technical Bulletin, Feb. 2021 [Online]. Available: https://multimedia.3m.com/mws/media/1791500O/comparison-ffp2-kn95-n95-filtering-facepiece-respirator-classes-tb.pdf

[51] A. Jacobs, "Counterfeit covid masks are still sold everywhere, despite misleading claims," *The New York Times*, Nov. 30, 2021. Accessed: Apr. 24, 2024 [Online]. Available: https://www.nytimes.com/2021/11/30/health/covid-masks-counterfeit-fake.html

[52] M. B. Westmoreland and T. Breton, "Amazon's three-pillar strategy to fight counterfeits," Jun. 15, 2022. Accessed: Apr. 24, 2024 [Online]. Available: https://www.aboutamazon.eu/news/policy/amazons-three-pillar-strategy-to-fight-counterfeits

[53] "Approximate conversions from U.S. customary measures to metric," *NIST*, Jan. 2010, Accessed: Jun. 08, 2024 [Online]. Available: https://www.nist.gov/pml/owm/approximate-conversions-us-customary-measures-metric

[54] Flour Advisory Bureau, "Baking conversion charts," Fab Flour. Accessed: Jun. 08, 2024 [Online]. Available: https://fabflour.co.uk/fab-baking/5-conversion-charts-for-your-convience/

# 2 Adhesives and Consolidants

## TWO-PART EPOXIES

Consultant: Astrid van Giffen

### *Material Information*

Precision Required: Medium

*Overview*

When walking down the fixatives aisle of your average hardware store, you might come across a large array of adhesives listed as 'epoxies'. In this general use, the term 'epoxy' can refer to any two-part, hard-setting adhesive system regardless of the chemicals involved.

In conservation the term is of course more specific, but even in this specialised context, epoxy retains some relative imprecision, referring to a broad range of epox*ide*-containing reactive (pre)polymers as well as their cured product. Epoxides are molecules which contain the eponymous functional group (Figure 2.1) consisting of cyclic ether, a three-atom ring made from two carbon and one oxygen. While compositionally only one small part of the larger molecules that comprise epoxy adhesives, this three-atom ring exhibits a characteristic energetic cross-linking behaviour that unites the otherwise diverse (pre) polymers.

Oxirane is an example of a solitary epoxide functional group, unincorporated in a larger molecule. As you can see in the image, the 60° bond angles in oxirane are far more acute than the 109° ideal for carbon. This cramped shape causes the ring to be under a great deal of strain, which in turn makes epoxides particularly reactive.

It is helpful to think of the epoxide ring as a kind of inverse mousetrap, ready to spring open at the slightest provocation. For the epoxy resins used in

DOI: 10.4324/9781003391708-3

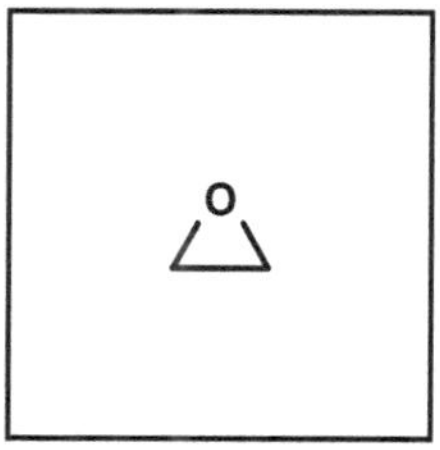

*Figure 2.1* Organic structure for oxirane.

conservation, this provocation is provided by the hardener (typically an amine), which prods the epoxide groups to swing open and bond with neighbouring atoms. This kick-starts a chain reaction, with epoxide groups breaking open and cross-linking with other polymer chains, eventually causing the entire system to become more rigid. This sudden vigorous cross-linking is the 'curing' process.

To stretch the analogy even further, imagine a small room of these epoxide mousetraps. Every square centimetre of the floor is covered with these tightly wound traps. Now imagine what would happen if you threw a handful of golf balls into that room. While there are only a few golf balls to drive the reaction, once started, it becomes self-perpetuating, the reactivity of the traps and their energetic change of shape driving further reactions.

In this image, the golf balls reflect the behaviour of the hardener, a small amount of which can incite a vigorous response. This, however, is where we must leave even this very loose analogy behind, for neither is the reaction as random and uncontrolled as this indicates, nor does excess hardener always cause faster curing. Instead, the ratio of hardener to epoxy resin is perhaps the most important variable in ensuring a correct cure.

The two-part epoxy systems discussed in this section are thus referred to in terms of their 'resin' and 'hardener' components. The epoxide-containing resin comprises the bulk of the adhesive mixture, while the hardener is the smaller precursor that catalyses the opening of the epoxide ring.

*Composition*

Most of the epoxy resins used in conservation are supplied as a two-part resin and hardener system. The epoxide-containing resins are often derived from a Bisphenol A base, which is then modified with various copolymers to produce a prepolymer with the desired characteristics. Hardeners typically consist of one of many different polyamine molecules, but phenols, anhydrides,

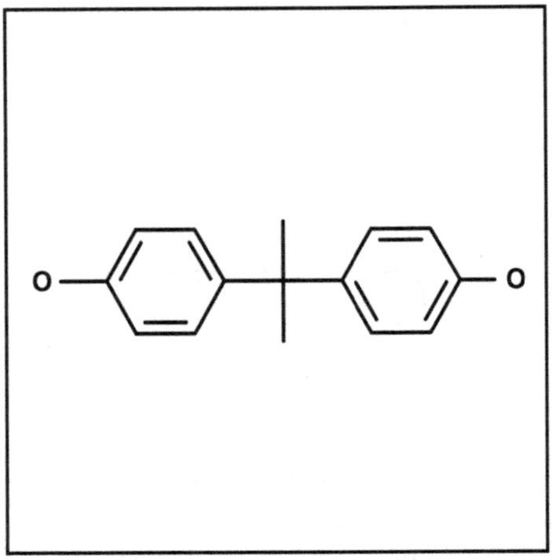


*Figure 2.2* Organic structure for Bisphenol A diglycidyl ether resin.

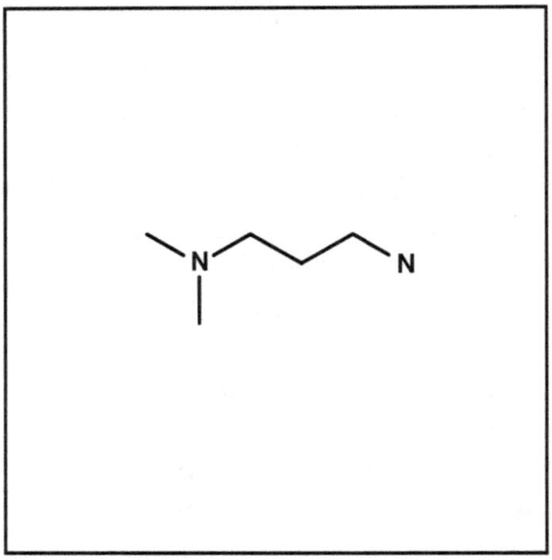


*Figure 2.3* Organic structure for N,n-Dimethylpropanediamine.

and thiols can also be used [1]. Although not discussed here, UV-curing epoxies are also common. The structures below illustrate two examples of the kinds of compounds used, but the exact composition of a particular epoxy is often complex and proprietary, and specific information is rarely freely available.

*Table 2.1* Collection of some useful ratios for common conservation epoxies

| *Epoxy* | *W:W ratio (resin: hardener)* | | | | | |
|---|---|---|---|---|---|---|
| Araldite 2020 | 100:30 | 1.0 g: 0.3 g | 2.0 g: 0.7 g | 5.0 g: 1.5 g | 7.0 g: 2.1 g | 10.0 g: 3.0 g |
| Hxtal NYL-1 | 3:1 | 1.5 g: 0.5 g | 3.0 g: 1.0 g | 4.5 g: 1.5 g | 6.0 g: 2.0 g | 9.0 g: 3.0 g |
| Epotek 301-2 | 100:35 | 1.0 g: 0.35 g | 2.0 g: 0.7 g | 4.0 g: 1.4 g | 6.0 g: 2.1 g | 10.0 g: 3.5 g |

### *Preparation: Epoxy*

#### *Techniques and Tips*

**Ratios**: According to manufacturer advice, maintaining the correct ratio of resin to hardener is the **most important** factor in epoxy preparation [1–4]. Before starting, see the sections in chapter 1 covering calibration and measurement. Be sure to work within the accuracy of your equipment. The information given in the technical section provides a good basis for investigation, but exact figures are liable to change suddenly and without notice. Always check manufacturer instructions for up-to-date information.

Many epoxies will provide only one ratio type (v/v or w/w) but wherever possible, *utilise weight-based ratios over volumetric ones*. This eliminates the need for separate measuring tools (pipettes, graduated cylinders, etc.), therefore reducing any variation or imprecision introduced in transferring materials from measurement to mixing containers (Table 2.1).

**Colouring**: Additives such as fumed silica or fumed titanium oxide can alter transparency and texture, but a broad range of colours can be achieved with the addition of various colourants. Purpose-made metal complex dyes (such as BASF Orasol®) produce the most homogenous colours, but good results can also be obtained with dry pigments, particularly those with extra-fine particles.

No matter the type used, the amount of colourant required is generally very small, and it takes time for it to fully dissolve into the resin. It's therefore recommended to add a small amount of colourant to the *resin component alone*, stir it well, cover it, and allow it to sit for several hours (ideally overnight). Hardener should be added only as the final step, once the desired colour has been reached [5].

Furthermore, to allow for greater adaptivity, several different coloured resins can be prepared at the same time, creating a colour palette for further mixing. These can be prepared as concentrated or highly saturated coloured resins, which can then be mixed, drop-wise, into clear resin. This enables more responsive colour-matching and allows several different colours and combinations of colours to be tested at the same time. This palette can also be stored (covered) and reused, although the exact viability duration is unknown [5].

Please note that epoxy must be allowed to fully cure before the colour match can be evaluated. The addition of hardener and/or the curing process itself can occasionally affect the result, in some cases drastically changing the colour. Finally, the addition of some dyes or pigments can also affect the setting time of the epoxy.

**Epoxy putties**: Epoxies can also be mixed with various bulking agents to produce a bespoke epoxy putty. To do so, simply add larger amounts of fumed silica, fumed titanium oxide, pigments, stone powders, and/or other aggregates until the desired consistency is reached. However, unlike colourants, these materials should be mixed into the resin only after the addition of the hardener. This lowers the working time but ensures that the reactive components are mixed homogenously.

**Curing times**: The cure time of an epoxy can be manipulated by strategically warming or cooling the resin mixture. For example, placing a pre-mixed epoxy in the freezer can slow the hardening process and extend its working life. Conversely, gentle heat can also help to speed up cure times, but this can have the additional effect of lowering the viscosity of the epoxy. Although not always desirable, this can have significant benefits for applications which rely on capillary action to draw adhesive into fine cracks. This is a temporary effect, however, and the longer you heat the epoxy, the more viscous it will become, and the more quickly it will stiffen upon cooling [5].

**Crystalized resin**: When stored at a too-low temperature, the resin part of some epoxies can crystalise. This is easily reversed by warming the sample until it melts. [5]

**Heat Release**: Epoxy curing is mildly exothermic, this can cause some mixing and moulding materials to distort, especially with larger amounts.

*Materials*

Balance (w:w), or two volumetric pipettes (v/v) (disposable is easiest)

Mixing container: watch glass, disposable weighing boat or plastic shot glass

Stirring rod (glass or bamboo) or scraper (silicone spatula or wooden tongue depressor). Avoid metal implements as these can cause staining or react with the hardener.

*Instructions*

Adapted from Koob (2004) [6]

Note: See the section on weighing and calibration.

1. Determine the amounts of resin and hardener required.
2. Measure out the resin and add the hardener. Although volumetric ratios might sometimes be necessary, ratios by weight (w:w) are strongly preferred for ease of preparation, accuracy, and quality of result.

   For w/w ratios: place the mixing cup on the balance and tare. Pour or pipette the resin into the container until you meet the desired mass. Tare the balance again. Using a separate clean pipette or the manufacturer-supplied dropper add very small amounts of hardener until you reach the desired mass. Go slowly; once added, the hardener can be difficult to remove.

   *Or*

   For volumetric ratios (v/v) – Use a clean pipette to measure out the desired volume of resin and add it to your mixing container. Be sure to minimize any residue left in the measure, squeezing or scraping out every last drop. Using a second clean pipette or the manufacturer-supplied dropper to decant the correct amount or hardener directly into the mixing container. Go slowly; once added the hardener cannot be removed.
3. Mix. Use a glass stirring rod or wooden tongue depressor to stir the epoxy mixture for one to two minutes. If possible, stir in a smooth figure-eight, avoiding incorporating air into the mixture. Bubbles are not only aesthetically displeasing for most applications, but excess air can also affect the curing process for some epoxies.
4. Rest slow-curing resins (such as Hxtal). While the resin mix can be used at this point directly, for best results set the mixture aside for five minutes, warm gently to about 35°C, and then stir it again for one full minute. This distributes any unused hardener within the resin, resulting in a more complete cure.

*Clean-up*

Materials which have come into contact with only either the resin or the hardener will not polymerize and can be wiped with paper towels, rinsed with solvent, and washed as usual. However, many, if not most, resins and hardeners are harmful to aquatic life. Avoid letting any residues into sinks or sewers and double-check the disposal information provided by the manufacturer.

For mixing containers, resin/hardener mixes can be wiped away as a liquid or allowed to partially cure to a semi-rigid gel before being picked out or pulled from the surface. Do not allow any resin mixtures to fully cure, as the hardened resin will bond with most surfaces and can ruin your glassware. Fully cured epoxy can sometimes be removed from glassware by soaking the glass in warm, soapy water or solvents for several hours or even days.

After all visible resin is removed, dampen a paper towel with acetone and wipe again to remove any remaining residues.

Immerse all materials in warm soapy water and wash normally.

Dispose of any wooden skewers or tongue depressors.

***Technical Information***

*Table 2.2* List of commonly used epoxy resins. However, *inclusion does not indicate recommendation for use on heritage materials*; many listed resins are semi-opaque or known to yellow [7].

| | *Resin: Hardener* | *Pot life* | *Ref index* | *Recommended cure* | *Minimum cure* | *Storage Temperature Expiry* | |
|---|---|---|---|---|---|---|---|
| **Araldite 2020** | w:w 100:30<br>v:v 100:35 | 40–50 min | 1.559 | None provided | 23 °C / 16 Hours<br>40 °C / 3 Hours | 18–25 °C<br>See container | [8] |
| **Araldite 2011**<br>***Araldite Standard*** | w:w 100:80<br>v:v 100:100 | 100 min | N/A - yellow | None provided | 23 °C / 7 Hours<br>40 °C / 2 Hours | 18–25 °C<br>See container | [9] |
| **Araldite 2012**<br>***Araldite Rapid*** | w:w 100:100<br>v:v 100:100 | 5–8 min | N/A - yellow | None Provided | 23 °C / <10 min | 18–25 °C<br>See container | [10] |
| **Epo-Tek 301** | W 20:5 | 1–2 Hours | 1.519 | 65 °C / 2 Hours | 65 °C / 1 Hours<br>23 °C / 24 Hours | 18–25 °C<br>1 year | [2] |
| **Epo-Tek 301-2** | W 100:35 | 8 Hours | 1.532 | 80 °C / 3 Hours | 23 °C / 48 Hours | 18–25 °C<br>1 year | [3] |
| **Hxtal NYL-1** | W 3:1 | None provided | 1.515 | 19 °C / 7 days | 19 °C / 3 days (limited movement) | 18–25 °C<br>No known expiry | [1] |

*Table 2.3* Composition information for a few commonly used epoxy resin – hardener systems

| | *Resin Composition* | *Hardener Composition* | |
|---|---|---|---|
| **Araldite 2020** | 50–62% diglycidyl ether of 1,4-butanediol | isophorone diamine | [11] |
| **Araldite 2011** **Araldite Standard** | 70–90% 2,2′-[(1-methylethylidene) bis(4,1-phenyleneoxymethylene)] bisoxirane 2.5–10% Formaldehyde, oligomeric reaction products with 1-chloro 2,3-epoxypropane and phenol 1–10% Bisphenol A – epoxy resins, number average MW >700 –<1100 | 5–9.65% N′-(3-aminopropyl)-N,Ndimethyl propane-1,3-diamine 3–5% Amines, polyethylenepoly-, triethylenetetramine fraction | [9] |
| **Araldite 2012** **Araldite Rapid** | 70–90% 2,2′-[(1-methylethylidene) bis(4,1-henyleneoxymethylene)] bisoxirane 3–10% 1,4-Bis(2,3-epoxypropoxy) butane | 2.5-10% 2,2′-[1,2- Ethanediylbis(oxy)] bis(ethanethiol 1-3% N′-(3-aminopropyl)-N,N dimethylpropane-1,3-diamine | [10] |
| **Epotek 301** | Bisphenol A diglycidyl ether | 2,2,4-Trimethylhexamethylenediamine | [7] |
| **Epotek 301-2** | Bisphenol A diglycidyl ether | Polyoxypropxylenediamine (>60%) | [3] |
| **Hxtal - NYL1** | Bisphenol A diglycidyl ether | triethylene diamine | [1] |

## RESINS INTO SOLVENTS

Consultant: Astrid van Giffen

### *Material Information*

Precision Required: Low-Medium

*Overview*

The term 'resin' describes a wide range of naturally occurring, naturally derived, and synthetically derived compounds. This can be chemically befuddling and is no doubt the reason that other publications prefer suitably specific discussions of polymer classes. However, for the purposes of this book, this lack of chemical specificity is usefully broad. For while the composition, chemistry, and use of these materials can vary enormously, the preparation of many resins used in conservation is often identical, comprising a simple dissolution: a solid solute mixed into a liquid solvent to a desired concentration.

This section will therefore focus on the preparation of solutions with specific *weight/volume* concentrations *and weight:volume* ratios. As discussed in greater detail in Chapter 1, in a working lab or studio, it is not always necessary for a solution to have a precise ratio of solvent to resin. It is, however, very important to understand the difference between these solution types, and why that difference matters.

Finally, these methods are for the dissolution of large solids (such as flakes or pellets) and not hydrophobic fine powders. Although many fine powders fall under the broad description of 'resins' these are largely covered in the section 'Cellulose Ethers'.

*Composition*

Despite what they may share in preparation and use, resins lack any significant compositional similarity. Resins include naturally occurring compounds (such as shellac), modified synthetic versions of naturally occurring compounds (cellulose acetate, nitrates, and ethers), as well as completely synthetic homopolymer and copolymer systems (Paraloids). Given this variety, and the pragmatic (lack of) scope of this text, those interested in specific composition information are referred to the table at the end of this section, the CAMEO website, or ideally C. V. Horie's excellent *Materials for Conservation* [12].

***Preparation: Resin into Solvent***

*Techniques and Tips*

**Concentrations vs Ratios**: As discussed in greater detail in Calculations Section in Chapter 1, weight-to-volume *concentrations* (w/v) are often conflated with weight-to-volume *ratio solutions* (w:v). While one system has no inherent functional benefit over the other, it is *vital* to understand that there is a difference. For example, a 40% dissolution of Paraloid B72 in acetone weight/volume concentration will have roughly *one-and-a-half-times* the amount of solids as a 4:10 weight:volume ratio.

**Solvent selection**: The solubility of polymers within various solvents has been described both in terms of Hansen parameters and Teas diagrams, and in the outright rejection of such methods [12, 13], Although this text won't retread such well-worn discussions, there are some common examples given in the technical data section for this recipe. It is worth mentioning that solvents can be combined to exploit different properties, such as slowing evaporation times. This only really works with miscible solvents, those that can completely dissolve into one another (ethanol and water) rather than non-miscible combinations (white spirit and ethanol).

Solvent choice not only affects the film properties and penetration of a given polymer; residual solvents have also been found to affect the glass transition temperature for Paraloid B44 and B72. Generally, heavier solvents (like toluene) which are slower to evaporate can remain in the Paraloid resin longer than smaller, quicker evaporating solvents like acetone, but all remnants can plasticise the polymer and depress the glass transition temperature [14]. This evaporation rate is influenced by vapour pressure and molecule size but has also been linked to polymer affinity and other factors. Warming can help drive the solvent from the polymer films, but this too can lead to slumping and loss of strength. Although further research is required, the setting time for Paraloid resins might need to be reconsidered. Those interested in more detailed discussions of this topic are encouraged to look up the Open Access article by Vinçotte et al. [14].

**Additives**: Like epoxies, most resins support the addition of colourants and bulking agents. Inert materials such as fumed silica, micro-balloons, fumed titanium oxide, and pigments can be added to provide additional viscosity or structure.

TOOLS: VOLUMETRIC CONTAINER

If more accurate w/vol concentrations are desired, then a method of measuring volumetric solutions is required. In a chemistry lab, this is generally accomplished with a volumetric flask, but the narrow neck that gives this glass its accuracy would frankly be an absolute nightmare to clean – not to mention the inflexibility of always making stock solutions. Instead, provided below is a simple method for turning your solution container into a volumetric measure. However, please note that while this method will give a passable approximation of a w/vol concentration, it is nowhere near accurate enough for anything requiring true precision.

*Materials*

Graduated cylinder
Well sealing container with lid
Permanent marker or wax pencil (anything that will mark on glass)
Desired solvents and resins

*Instructions*

1. Decide the total amount of solution you'd like to prepare. For example, 30 mL of 15% w/v.
2. Fill the graduated cylinder with tap water to the desired total amount, in this example 30 mL.
3. Pour the water into your container and mark the level of the water with the marker. Ensure that the bottom of the mark aligns with the bottom of the meniscus.

4. Empty the water from the container and dry carefully. Ensure that all water is removed but that the mark you made is still visible. A solvent rinse might be necessary to remove trace moisture.
5. Create a resin-solvent solution using one of the techniques outlined below.

*Method: Tea Bag*

This method is suitable for higher concentrations or slow-dissolving systems and is a preferred method for dissolving resin pellets. This recipe was adapted from the system described by S. Koob [15].

MATERIALS

Container with lid (volumetric optional)
Balance
Polyester netting
Fishing line or thread
Scissors
Resin of choice
Solvent of choice
PPE and extraction as necessary

INSTRUCTIONS

1. Create a volumetric container or simply use a small container with a tight-fitting lid.
2. Weigh out the desired amount of resin solids.
3. Wrap the polyester netting around the resin pellets and tie off with the thread, essentially creating a resin 'tea bag'. Keep an eye on the width of the container, ensuring that the resin parcel will fit through the mouth and neck. Lastly, when tying off the bag, leave one or two of the threads long; this will be used to suspend your tea bag within the solvent in a later step.
4. Take your lidded container (DIY if applicable) and add the solvent. If preparing a w/v concentration, fill the container roughly halfway with the desired solvent. The solvent:resin ratio needs to be high enough to encourage dissolution but should fall well short of your desired final volume. If preparing a w:v ratio, feel free to add the full amount of solvent.
5. Holding the tea bag by the long threads, suspend the bag in the container so that only the very bottom of the tea bag barely touches the solvent line. As the resin dissolves and sinks, the solvent with lower polymer load will sit at the top. Keeping the bag relatively high keeps it in direct contact with this less-concentrated solvent, encouraging quicker dissolution.
6. Either tape the long threads to the outside of the container or pinch them in place using the lid. Close the container tightly.

7. Allow to sit until tea bag is empty and the resin has fully dissolved into the solvent. You may need to adjust the height of the bag as the solution rises to ensure only the very bottom stays in contact with the solvent. Depending on the resin, solvent, concentration, and other factors, this can take from several hours to several days.
8. If using a volumetric container, add solvent until the level reaches the final volume mark.
9. Mix in any desired additives, such as fumed silica or pigment. Stir well.
10. Seal and label.

*Method: Leave-It*

As the name indicates, this isn't exactly a process, but is in common enough use to merit a mention. Solutions with low concentrations and systems where the resin readily dissolves in the desired solvent can be prepared much more simply. In these instances, the desired ratio can simply be tipped in a container, sealed, stirred through mild agitation or swirling, and then left overnight.

*Clean-up*

After solution preparation, most implements can be simply wiped with paper towels or disposed of directly. However, removing old, water-insoluble resin from glassware is worth a short mention.

1. Pour or pick out any undesired resin onto a paper towel or piece of scrap paper.
2. Place this in a well-ventilated space or fume cupboard to allow solvent to evaporate. Once dried, dispose of hardened polymer according to manufacturer instructions.
3. To remove residue from glassware: place a small amount of solvent in the container.
4. Seal and swirl the solvent until the remaining resin has dissolved. This can take anything from several hours to a few days, but overnight will suffice for most.
5. Once dissolved, add a fair amount of water to the solvent. Water insoluble synthetic resins will begin to precipitate, becoming stringy, sticky, and water white. A skewer or gloved finger swirled in the container will quickly collect the polymer threads from the surface of the glassware.
6. Glassware can then be soaked in warm, soapy water and washed as usual.

***Technical Data***

Please note that how readily a listed polymer will actually dissolve in the reported solvents can and will vary. Solvents anecdotally reported to enable quick dissolution are indicated in **bold**, but this is neither conclusive nor comprehensive.

Please note that the chemical composition and glass transition temperature listed reflect those supplied by the manufacturer. Previous investigations have noted variations between such reported figures and those determined experimentally [16, 17] (Table 2.4).

*Table 2.4* Collection of commonly used resins and some properties.

| *Trade name* | *Chemical composition* | *Tg (°C)* | *Concentrations (w/v)* | *MW* | *Solvents* | *Source* |
|---|---|---|---|---|---|---|
| Aquazol 50 | poly(2-ethyl-2-oxazoline) | 69–71 | 10–20% in water or alcohol for consolidation. | 50,000 | Acetone, ethanol, methanol, **water**, propylene glycol, methyl ethyl ketone | [18–20] |
| Aquazol 200 | poly(2-ethyl-2-oxazoline) | 69–71 | 2.5% in propan-2-ol consolidant<br>10–50% in water or alcohol for adhesion. | 200,000 | Acetone, ethanol, methanol, **water**, propylene glycol, methyl ethyl ketone | [18–21] |
| Aquazol 500 | poly(2-ethyl-2-oxazoline) | 69–71 | 20–50% adhesion | 500,000 | Acetone, ethanol, methanol, **water**, propylene glycol, methyl ethyl ketone | [18–21] |
| Butvar B-76 | Polyvinyl Alcohol 11.5–13.5%<br>Polyvinyl acetate 0–2.5%<br>Polyvinyl butyral 88% | 62–72 | Up to 5% in ethanol consolidation<br>20% for adhesion (>25% and too bulky) | 90,000–120,000 | Acetone, DMSO, isopropyl alcohol, 95% or anhydrous ketones*, alcohols*, toluene/ethyl alcohol, 95% (60:40 by weight), THF, methanol (partially) | [22, 23] |
| Butvar B-98 | Polyvinyl Alcohol 18–20%<br>Polyvinyl acetate 0–2.5%<br>Polyvinyl butyral 80% | 72–78 | 5% in IMS consolidation<br>5% IMS/toluene 1:1<br>10% acetone consolidation<br>20% IMS adhesive | 40,000–70,000 | Acetone (partially in doc))<br>DMSO<br>Isopropyl alcohol, 95% or anhydrous<br>Toluene/ethyl alcohol, 95% (60:40 by weight)<br>Methanol<br>THF | [16] |
| Synthomer M30* | Polyvinyl acetate | 36 | <10% consolidant<br>20–40% adhesive | 85,000 | Many esters, ketones, aromatic hydrocarbons and lower alcohols<br>**Ethanol** | [24] |

| | | | | | | |
|---|---|---|---|---|---|---|
| Synthomer M40* | Polyvinyl acetate | 40 | <10% consolidant<br>20–40% adhesive | 120,000 | Many esters, ketones, aromatic hydrocarbons and lower alcohols<br>**Ethanol** | [25] |
| Synthomer M50* | Polyvinyl acetate | 42 | <10% consolidant<br>20–40% adhesive | 176,000 | Many esters, ketones, aromatic hydrocarbons and lower alcohols<br>**Ethanol** | [26] |
| Mowital B20H | Polyvinyl butyral 75–81%<br>Polyvinyl alcohol 18–20%<br>Polyvinyl acetate 1–4% | 64 | <10% consolidant<br>20–40% adhesive | 17,500–27,500 | Acetone, **Ethanol**, butanol, propanol, diacetone alcohol, glycol ethers, cellusolve™, methylethylketone, DMSO, THF | [27] |
| Mowital B30H | Polyvinyl butyral 75–81%<br>Polyvinyl alcohol 18–20%<br>Polyvinyl acetate 1–4% | 68 | <10% consolidant<br>20–40% adhesive | 27,500–37,500 | Acetone, **Ethanol**, butanol, propanol, diacetone alcohol, glycol ethers, cellusolve™, methylethylketone, DMSO, THF | [27] |
| Mowital B60H | Polyvinyl butyral 75–81%<br>Polyvinyl alcohol 18–20%<br>Polyvinyl acetate 1–4% | 70 | <10% consolidant<br>20–40% adhesive | 50,000–60,000 | Acetone, **Ethanol**, butanol, propanol, diacetone alcohol, glycol ethers, cellusolve™, methylethylketone, DMSO, THF | [27] |
| Paraloid B-44 | Methyl methacrylate:ethyl acrylate:n-butyl methacrylate 70:28:1 (H) | 60 | 20–40% surface coating | 105,000 | **toluene, xylene**, MEK, and some esters and glycol ethers (Dow) but insoluble in mineral spirits | [12, 28] |

(*Continued*)

*Table 2.4* (Continued)

| *Trade name* | *Chemical composition* | *Tg (°C)* | *Concentrations (w/v)* | *MW* | *Solvents* | *Source* |
|---|---|---|---|---|---|---|
| Paraloid B-48 N | Methyl methacrylate Butyl acrylate copoolymer | 50 | 15–30% adhesive | 250,000 | **toluene, xylene**, MEK, and some esters and glycol ethers but insoluble in mineral spirits | [12, 29] |
| Paraloid B-67 | iso-Butyl Methacrylate | 50 | 10–20% surface coating | 60,000 | **toluene, xylene**, ethyl acetate, VM&P Naphtha, mineral spirits, acetone, MEK | [12, 30] |
| Paraloid B-72 | ethyl methacrylate 70% methyl acrylate 30% copolymer. | 40 | 5–10% consolidant<br>20–50% adhesive<br>30–40% casting | 105,000 | **Acetone**, **toluene**, xylene, MEK, ethanol | [12, 31] |
| Paraloid F-10 | BMA copolymer | 20<br>>38 tacky surface | n/a | 150,000 | Supplied as 40% solids in blend of mineral spirits | [12, 32] |

* Newly renamed Mowilith range.

## CELLULOSE ETHERS

Consultant: Mathilde Renauld

***Material Information***

Precision Required: Low-Medium

*Overview*

Cellulose ethers (CE) can be used in a range of applications, from gels and adhesives to paper sizing. Unlike the broad categorization of previous sections, the preparation of many cellulose ether solutions shares certain techniques which are unique to this class of materials. If prepared using the general methods outlined in 'Resins', the fine CE powder will typically quickly clump together, forming insoluble lumps that can fail to dissolve completely.

However, while there are overlapping approaches and characteristics between types of CE, the preparations can vary in significant ways. It is therefore important to ensure that the correct technique is applied for each specific cellulose ether/solvent combination. The recipes in this section provide guidance for which approach is appropriate for each polymer type.

Lastly, after belabouring the difference between w/v and w:v ratios in previous sections, for most bench work, there is no significant benefit to making true w/v volumetric solutions for cellulose ethers. The concentrations used in these preparations are considered too low for the density difference between solute and solvent to affect the resulting solution. Therefore, most solutions with w:v ratios approximate w/v concentrations and for all but the most precise uses, this is sufficient.

*Composition*

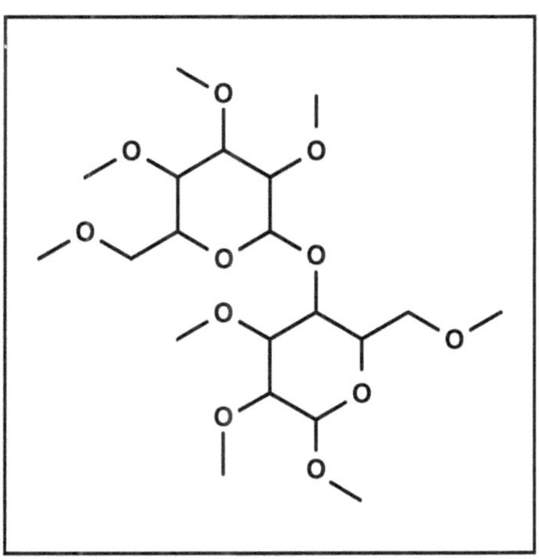

*Figure 2.4* Segment of repeating organic structure for methyl cellulose.

A small section of methyl cellulose polysaccharide is shown to highlight the alternating conformation of ether bridges. Although the branching and substitution can and does vary, this demonstrates the basic backbone of cellulose ether polymers. The location, type and degree of substitution of the hydroxyl groups is what gives various cellulose ethers their specific properties.

***Preparation: Cellulose Ethers***

All recipes adapted from Feller and Witt 1990 [33] and Ashland Manufacturing [34]

*Method: "Slurry"*

Polymers: Methylcellulose (MC), Hydroxypropyl Cellulose (HPC), Hydroxypropyl Methylcellulose (HPMC)

Note: Only really appropriate for aqueous preparations, if using alcohol or other solvents see the 'shake' and 'stir' methods.

MATERIALS

Cellulose ether powder and deionized water
Hot plate with magnetic stirrer, or hot plate and small whisk.
Balance
Thermometer: desirable for Hydroxypropyl Cellulose (HPC) but optional for other CEs
Graduated cylinder
Beaker
Heat-proof container with lid: this should have a volume that is greater than the total amount of water.
Watch glass or weighing paper
Heat-proof gloves
Heat-proof mat or small ceramic tile (optional)

INSTRUCTIONS

1. Label the heat-proof lidded container with the materials and concentrations.
2. Use the graduated cylinder to measure out the total desired amount of water.
3. Using the watch glass or weighing paper, weigh out the desired amount of solids and set aside. CEs are typically very light and can be easily scattered by air flow or swift movement. Cover with an extra watch glass if required.
4. Pour roughly 1/3 of the distilled water into the heatproof glass container and place the container on the hot plate. The amount of water used for this hydration step is flexible and only needs to be a minimum of 3–8x the volume of the CE used. A third usually more than covers it. Place the container's lid to one side for now.

5. Cool the remaining (2/3) water by placing the graduated cylinder or beaker into a refrigerator (if available) or simply set aside in a cool area.
6. Turn on the hot plate to a low or medium setting and warm the water in the heat-proof container. Heat the water until you reach the dispersion temperature listed in Technical Information. For many CEs a point above steaming but below a simmer is sufficient, but you can use a thermometer for greater accuracy. A thermometer is suggested for preparations of hydroxypropyl cellulose (HPC) given its relatively narrow dispersion range.
7. Using a magnetic stir bar or whisk, stir the warm water forming a consistent but not overly vigorous vortex. This extra agitation can initially cool the water slightly so further adjustments to temperature may be required. Slowly add the CE powder to the inner wall of the vortex and mix until the powder is evenly dispersed and fully hydrated. Take your time. This process usually takes several minutes, but lower concentrations will prepare more quickly than higher.
8. Stop stirring and heating. If using a combination hot plate–stirrer, simply turn off both components and allow the mixture to cool slightly. If hand stirring and the height of the hot plate is awkward, you can turn off the heat and remove the warm slurry from the hot plate completely. If doing so, use the heat-proof gloves and place the warm container on a small heat-proof mat, cotton towel, or ceramic plate. It's worth remembering that 70°C can scald skin within seconds.
9. Remove the remaining cool water from the refrigerator.
10. Stir the warm CE-water mixture again and add the remaining cool water to the hot slurry. Continue stirring until well-incorporated and set aside.
11. The solution will thicken as it cools, taking anywhere from 20 minutes to several hours. Once the container is no longer hot, seal lid.

*Method: Shake and Sit*

Polymers: Methylcellulose (MC), Hydroxypropyl Cellulose (HPC), Hydroxypropyl Methylcellulose (HPMC), Ethylhydroxyethyl Cellulose (EHEC), Sodium Carboxymethyl Cellulose (NaCMC)

Note: Particularly at higher concentrations, this method can result in a lumpy, uneven solution.

MATERIALS

Cellulose ether powder
Desired solvent
Container with tight-fitting lid
Graduated cylinder
Beaker
Watch glass or weighing paper

INSTRUCTIONS

1. Label the lidded container with the materials and concentrations.
2. Use the graduated cylinder to measure out the total desired amount of solvent.
3. Using the watch glass or weighing paper, weigh out the desired amount of solids and set aside. CEs are typically very light and can be easily scattered by air flow or swift movement. Cover with an extra watch glass if required.
4. Pour roughly half of the solvent into the lidded container.
5. Add the solute to the solvent, seal and shake for at least one minute.
6. Allow solution to rest for a few minutes.
7. Add the rest of the solvent, seal and shake again.
8. Visually inspect mixture for any clumps of CE power; break these up with a gloved finger or implement.
9. Set aside to continue to thicken.

*Method: Stir and Sit*

Polymers: Methylcellulose (MC), Hydroxypropyl Cellulose (HPC), Ethylhydroxyethyl Cellulose (EHEC), Sodium Carboxymethyl Cellulose (NaCMC)

MATERIALS

Cellulose ether powder
Desired solvent
Container with lid
Graduated cylinder
Beaker
Watch glass or weighing paper
Stirrer: whisk or magnetic stir bar and stirrer

INSTRUCTIONS

1. Label the lidded container with the materials and concentrations.
2. Use the graduated cylinder to measure out the total desired amount of solvent.
3. Using the watch glass or weighing paper, weigh out the desired amount of solids and set aside. CEs are typically very light and can be easily scattered by air flow or swift movement. Cover with an extra watch glass if required.
4. Pour all solvent into the lidded container.
5. Using a magnetic stir bar or whisk, stir the solvent, creating a gentle vortex. Slowly add the CE powder and mix until the powder is evenly dispersed.

6. Visually inspect mixture for any clumps of CE power; break these up with a gloved finger or implement.
7. Set aside to continue to thicken.

***Storage***

Regardless of preparation type, the solution will precipitate and settle over time. Shake or swirl the container a few hours before use. Cellulosic materials are also food sources for bacteria. Consider storing solutions in a refrigerator and dispose if discolouration or growths are noted.

***Technical Data***

*Table 2.5* Collection of cellulose ethers and some properties.

| | *Dispersion Temp (°C)* | *Concentrations* | *Solvents* | *Trade names* | |
|---|---|---|---|---|---|
| Methylcellulose (MC) | 70<br>80–90 | 0.5–5%<br>8–10% | Room temperature and cool water (insoluble above 45 °C) | Methocel A4C (41,000)<br>Methocel A4M (86,000) | [33, 35] |
| Hydroxypropyl Cellulose (HPC) | 55 +- 5<br>Not above 60 | 0.5–5% for most applications | Room temperature and cool water (insoluble above 45 °C)<br>Organic polar solvents: methanol, isopropanol, acetone, DMSO | Klucel G (300,000)<br>Klucel E (60,000) | [33–35] |
| Ethylhydroxyethyl Cellulose (EHEC) | 70 | 2–4% consolidant, adhesive<br>8–10% gel | Cold water, can be thinned with alcohol or acetone. | Ethulose<br>Bermocoll E 481 FQ | [33, 35] |
| Hydroxypropyl methylcellulose (HPMC) (Hypromellose) | 70 | 1–5% | Cold water | | [33, 35] |
| Sodium Carboxymethyl cellulose (NaCMC) | 70 | 0.5–3% | Hot or cold water | Cellofas B-3500<br>Ashland Aqualon CMC | [33, 35] |

## STARCH PASTE

Consultant: Mathilde Renauld

Note: Preparing starch paste is a multi-stage process; read through all sections (preparation, cooking, storage, sieving) before beginning the cooking process.

### ***Material Information***

Precision Required: Low

*Overview*

Naturally occurring starches are found throughout the plant kingdom and their adhesive properties have been exploited for millennia. In particular, the scroll-mounting traditions in Japan and China have a long history of preparing highly refined starch pastes, some of which can take decades to produce. This knowledge is often shared through lengthy apprenticeships and is something to be slowly mastered rather than quickly taught [36, 37]. By all accounts these sophisticated techniques produce adhesives with characteristics that cannot be replicated with faster and easier means. However, the typical constraints of time and resources have also led to the development of much quicker, but still effective, preparation techniques.

Perhaps as a side effect of these conflicting legacies, there are numerous ways to make starch paste ranging from slow and steady to zapping in a microwave. The literature on this is enormous, nuanced, and broadly available. The 'BPG Recipes' Wiki for the Book and Paper Group of the American Institute for Conservation and Matsumaru's excellent review (2021) both provide excellent entry points for the non-specialist conservator [38, 39].

The starch pastes used in conservation work are typically derived from rice or wheat (although tapioca and arrowroot starch pastes are also in use). Of the two main sources, wheat starches are best known and are further described as either 'fresh' or 'aged'. The general term 'wheat starch paste' (WSP), most often refers to the former, 'fresh' paste, although this material is also sometimes known by the Japanese terms *shinnori* or *shōfunori*. Conversely, 'aged' WSP is known almost exclusively by the Japanese term *furunori* and requires highly specialised preparation.

**Rice starch**: Starch derived from rice is also available although less widely used. Lighter and thinner than WSP, RSP makes a much weaker adhesive with a neutral pH and much lower tendency towards gel formation [12].

**(Fresh) wheat starch paste (WSP)**: the most commonly used starch adhesive. WSP can be derived straight from flour for non-archival projects such as paper craft or fly postering. Conservation-grade starch is purified and highly refined, resulting in granules of a particular size, and is available from conservation suppliers under the proprietary name Aytex-P or the general term *zen/jin shōfu*

***Furunori***: Although far too specialised to be covered in this volume, *furunori* is an aged WSP that is historically and concurrently used in scroll mounting.

However, even among starch pastes, its preparation is complex, specialised, difficult to replicate, and the subject of significant discussion. It is noted for completeness' sake and to distinguish it from *funori*, a seaweed-derived adhesive discussed in in the section on seaweed gums [36].

While the needs of a given conservation project will always dictate the desired consistency of any adhesive mixture, WSP is particularly subjective and adaptable. The measurements given provide a loose starting point for a basic paste which will then be further adapted prior to use. As post-cooking dilution is simpler than concentration, the recipes given below produce a slightly thicker paste, likely in need of further thinning prior to use.

However, this is not to say that all starch recipes will produce essentially the same paste. While it will not be discussed here, it is worth reiterating the significant volume and depth of literature on the topic of WSP in conservation. Both subjective accounts and qualitative measurement describe how the starting mixture, soaking times, and the timing, heating, and mixing of the cooking process will affect the properties of the paste.

*Composition*

Starch occurs in nature as small water-insoluble granules produced as an energy storage method for green plants. Starch is primarily a mix of linear amylose and branched amylopectin molecules, but the exact proportions of these will vary based on the source of starch (Technical Information). Like cellulose ethers, the base unit of these two polymer chains is a D-glucose ring, but here the two types are differentiated not by their hydroxyl substitutions, but by small differences in bonding that results in large changes in structural architecture.

Before being used as an adhesive, these granules must be swollen and heated until they burst, allowing the amylose and amylopectin molecules to uncoil, disperse, and interact. As the mixture cools, the amylose molecules associate into double helical structures, cross-linking to form a gel [12].

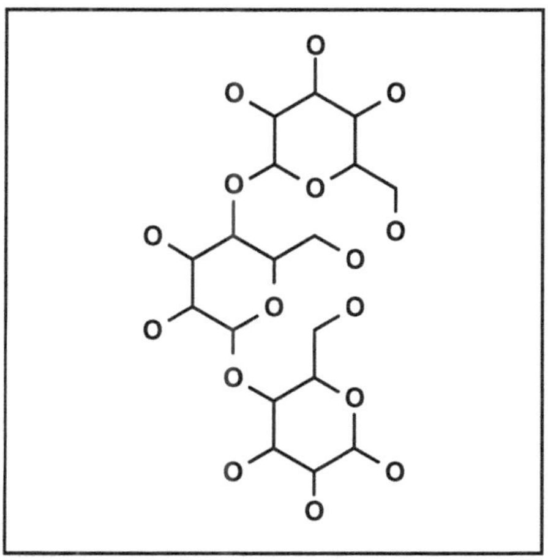


*Figure 2.5* Organic structure for amylose.

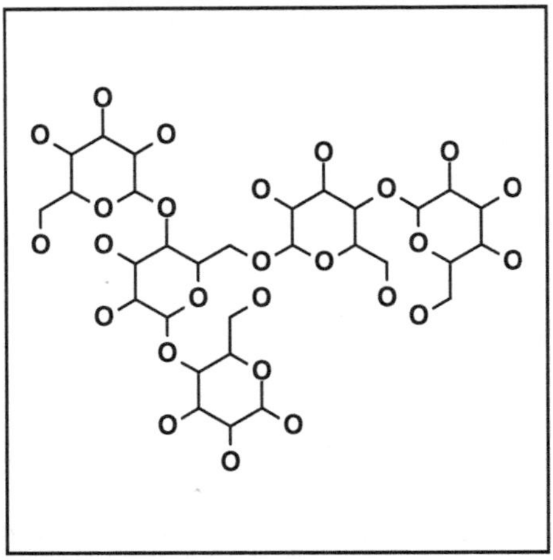

*Figure 2.6* Organic structure for amylopectin.

***Starch Paste Storage***

Unfortunately, even if refrigerated, starch pastes are quick to rot, sprouting mould after only a few days. Doubly unfortunately, while refrigeration is thought to discourage mould growth, cool temperatures can encourage a process called retrogradation. Here the amylose and amylopectin molecules begin to recrystalise, essentially transforming the flexible starch network back into a more structured, rigid form.

For these reasons, starch pastes are typically prepared in small batches a few hours or a day before their use. Storage methods vary, and only a small sample of techniques is included below, but all balance the risk of mould and retrogradation in a way that favours one deterioration mechanism over the other.

For those hoping to use a larger amount of paste more quickly, the *water-covered* method requires less preparation, but the pastes often mould after a few days.

Those using very small amounts of paste either in situ, on site, or infrequently should see the following *sterile syringe* method. This method allows for very long-term storage without risk of mould, but requires some preparation and can only dispense a small amount of starch paste.

*Storage Methods*

WATER-COVERED

1. Before cooking your paste, fill a clean (ideally sterile) storage container half-full with water (tap is fine in most contexts) and place into the refrigerator to cool. The container should have ample space for both the paste and some water.

2. Immediately following cooking, remove the container from the refrigerator and scrape the warm paste directly into the cool water. Keep an eye for any displaced water spilling out of your container but feel free to simply tip out any excess. The water should fully cover the paste.
3. Once cool, the paste can be returned to the refrigerator for later use but is ready to be kneaded, sieved, and used immediately.

STERILE SYRINGE

Miller and Root found that decanting the hot paste directly into sterile syringes allows the paste to be kept at room temperature for several months without growing mould [40]. These results were easily replicated by the author, and while the paste was eventually found to dry and shrink, no mould was observed within the six-month storage duration. All this said, the overlong periods used to assess this method have implications for retrogradation and changes in properties, and such extended storage should be avoided.

1. Pull a new sterile syringe from its wrapper and label. All syringes used in testing were new and sterile, but with proper cleaning, the syringes should be able to be reused or at least repurposed.
2. Immediately following gelatinization, while the paste is still hot and fluid, immerse the tip of the syringe into the paste and draw it up into the barrel of the syringe.
3. Squeeze out a small amount of paste to remove any bubbles.
4. Dampen a tissue with alcohol (methylated is fine) and wipe the excess paste from the tip.
5. Cap the syringe using either the plastic lid provided, or by using the plastic base of a disposable needle (snapping off the needle before use).
6. Store in a dark cupboard until needed.

***Preparation: Wheat Starch Paste (WSP)***

Starch derived from wheat boasts both a long history of increasing refinement and use, as well as impromptu use in all manner of home projects. It is worth noting that the recipes given below will also produce a workable paste by simply using store-bought, bog-standard, unpurified wheat flour. However, while the resulting wheat paste can be helpful for arts practice or home projects, the gluten will degrade faster than the starch, eventually cross-linking and becoming irreversible. The instability of the gluten is why conservation-grade adhesives are typically listed as 'gluten-free', 'jin/zen shofu', 'refined', or 'highly refined', indicating that the starch granules have been isolated from the gluten and other molecules found in flour.

This is a multi-step process taking about 30–45 minutes to complete. Ensure you read **all** instructions on Storage, Cooking, Kneading, Sieving, and Dilution prior to collecting materials.

*Techniques*

As with everything to do with WSP there is a refinement range based on preference and quality of the paste desired. Here are a few quick notes.

**Pre-soaking**: Soaking the starch powder prior to cooking (anywhere from 30 minutes to 10 hours) is thought to reduce the end cook time and yield a more consistent paste. The mechanism or efficacy of this has yet to be formally determined and it is not strictly necessary [38].

**Stirring**: Stirring while cooking should be quick and consistent without being overly vigorous. The aim is to ensure even cooking and is not dissimilar to making a béchamel sauce. The pace should be somewhere around 60–80 rpm, which, if you need a memory prompt, corresponds roughly to the beat of 'beyond the sea' as sung by Bobby Darin. As the paste thickens and begins to gel, quicker, more vigorous stirring will be required to keep it from overcooking and will help to burst the swollen granules.

**Cooking**: Unfortunately for the non-specialist conservator, there is no established cooking duration for WSP; the paste is simply done when it's done. This is typically determined by keeping a close eye on the changing characteristics of the paste as it cooks. The mixture goes from a thin, opaque white, gradually gelling, becoming glossier and more translucent. You will also notice a light aroma as the starch begins to cook, perhaps not as nice as béchamel, but not unpleasantly bready. There are several excellent YouTube videos which depict these changes, the time lapse on the AIC&FAIC channel being a particularly good example [41].

*Tools*

**Scoop** – Any volumetric measurement will work so long as the portions are measured consistently. Examples include: tablespoons (see Imperial Conversion Table 1.8); 30 mL scoops; or small glass beakers.

**Cooking pot** – Any small, affordably priced saucepan will do, but if possible, avoid those with an aluminium or copper cooking surface. It is thought that these can cause reactive ions to enter the paste [42]. Non-stick works very well.

**Wooden stirrer** – A straight rolling pin (with a diameter over 1 cm) is preferred by some, but a bamboo rice paddle can also work well. The square-headed ones in particular are good for getting into the corners of the pan. For the microwave method, a dowel rod was found to be an ideal tool for the smaller batches.

**Sieve/strainer** – Japanese horsehair sieves reportedly yield the smoothest product, but can be prohibitively expensive and their use requires some finesse. For the non-specialist conservator, a nylon screen (such as those used in printmaking) or fine metal tea strainer will also work well if the project can tolerate the presence of any metal particles that may flake off [39].

*Method: Hot Plate*

MATERIALS

Refined wheat starch powder: preferably conservation grade
Water: distillation may be required for some projects but tap water will work fine in most contexts.
Scoop
A cooking pot
Wooden stirrer
Storage container
Fine sieve

INSTRUCTIONS

Starting ratio: 5 parts water to 1 part dried starch (v:v)

1. Using a level scoop, measure out the dried starch and add it to the cooking pot. Using the same measuring scoop, portion out the desired amount of water, add the water to the starch and mix well.
2. *Optional*: Let the starch mixture sit for anywhere from 30 minutes to overnight, thoroughly wetting the granules. Mix again immediately prior to cooking.
3. Once the water and starch are well mixed, begin heating the mixture (medium-low to medium depending on your device), and continue stirring regularly. The paste will begin to thicken, but do not allow it to boil. Temperatures should ideally reach the range of 58–64°C (136.4–147.2°F). Cook for a total of 20–25 minutes, stirring constantly.
4. The mixture will change its consistency and appearance several times throughout the cooking process. By the end, the paste should be smooth, translucent, and roughly the consistency of jam, clinging to the stirrer and dripping slowly back into the pot, not running freely. If lightly overcooked to a thicker consistency, the paste can still be used, but will require additional straining and thinning.
5. Once cooking is complete, remove the pan from the heat and decant the paste into the desired storage method (see notes on storage). Decant while warm. The paste will thicken as it cools.
6. Sieve and thin to desired viscosity immediately prior to use. See following section on 'sieving and diluting'.

*Method: Microwave*

MATERIALS

Refined wheat starch powder
Distilled or deionized water

Scoop

Heat-proof large glass beaker. It is worth noting that the mixture will swell anywhere from *two to five times its volume* during cooking. Ensure you select a vessel that allows space for this expansion.

Towel or oven mitt: strictly necessary. The paste will get very hot during cooking and can quickly scald.

Small wooden stirrer or glass stirring rod

Microwave: ideally with working interior light

Storage container: see 'Storage' section below

#### INSTRUCTIONS

Note: This method is a bit rough and ready but will certainly produce a perfectly workable paste. Close attention should be paid to stirring the mixture during cooking to agitate the granules and encourage even heating.

Starting Ratio: 1 part dried starch: 6 parts water (v/v)

1. Add the water to the starch and mix.
2. Set the power level of the microwave to the lowest available setting. Remember you are aiming to cook the paste without boiling or burning it. This requires careful observation and short bursts of cooking followed by intense stirring. The beaker will get very hot during the cooking process, you will need to use something to allow you to safely handle the beaker such as an oven mitt or several layers of folded paper towel.
3. Cook the mixture for up to 20 seconds. Remove from microwave and stir vigorously. You need to break up any lumps as they form and mix in any uncooked starch granules.
4. Return to the microwave, heating for another 10–30 seconds and then stirring again. Each time, the mixture will quickly swell, bubble, and begin to boil. Do not allow it to bubble more than a few seconds before stopping the microwave to stir again.
5. Repeat this another 6–10 times, resulting in a total cook time of between 2 and 5 minutes (depending on batch size and microwave power). The mixture will change its consistency and appearance several times throughout the cooking process. By the end, the paste should be smooth, translucent, and roughly the consistency of jam, clinging to the stirrer and dripping slowly back into the pot, not running freely. If lightly overcooked to a thicker consistency, the paste can still be used, but will require additional straining and thinning.
6. Once cooking is complete, decant the paste into the desired storage method (see 'Storage Methods'). Decant while warm. The paste will thicken as it cools.
7. Sieve and thin to desired viscosity immediately prior to use. See 'Sieving and Diluting'.

### *Preparation: Rice Starch Paste*

The preparation of rice starch paste, despite its compositional similarity to WSP, is considerably less complex, with much less literature devoted to the topic.

#### *Materials*

Rice starch powder (preferably conservation grade)
Distilled or deionized water
Balance
Graduated cylinder
Watch glass or weighing paper
2 glass beakers or one beaker and one thick-bottomed metal pot.
Wooden stirrer (such as a dowel rod) or glass stirring rod.
Hot plate
Microwave
Storage container (see Storage)
Fine sieve (see Sieving)

#### *Instructions*

Ratios: 1 g rice flour: 10 mL cold water: 15 mL hot water

1. Using the watch glass or paper, weigh out the desired amount of rice flour and add to a glass beaker.
2. Use the graduated cylinder to measure out the volume of cool water required and add to the beaker with the rice flour. Stir well until no lumps remain.
3. Use the graduated cylinder to measure out the volume of hot water required and add to the second beaker or pot. Place this on the hot plate and bring to a gentle boil.
4. Once boiling gently, stir the hot water creating a vortex.
5. Continue stirring and slowly add the starch/cold water slurry to the hot water in a steady stream.
6. Once fully incorporated, stir constantly and continue to cook at the same temperature for a few minutes.
7. Once cooking is complete, decant the paste into the desired storage method (see 'Storage Methods'). Decant while warm. The paste will thicken as it cools.
8. Sieve and thin to desired viscosity immediately prior to use.

### *Final Steps*

No matter how the starch paste is cooked or stored, it is usually too lumpy, thick, and uneven to be usefully applied. As the final steps taken immediately prior to use, all pastes must be sieved and/or kneaded to produce a smooth workable paste.

*Kneading*

Kneading is an optional process that improves the paste's consistency. There are specialist bowls and sieves for this. As it is not strictly necessary, it is not included, but interested parties are referred to Mito Matsumaru's 2022 blog post [43].

*Sieving and Diluting*

This is the final step in the process and is completed immediately prior to use. Sieving is what turns the thick starch into a usable adhesive, but as it further breaks down the starch, it cannot be done in advance. As with everything WSP-related, there are countless approaches and nuances to preparing the paste for use. Again, see Matsumaru 2022 for information on correct use of horsehair sieves [43]. What all approaches share, however, is that the consistency of the paste should be refined by repeatedly pressing it through a fine screen.

A horsehair sieve is traditional, a nylon screen (like those used in screen printing) is also acceptable, and a tea strainer will do in a pinch (if trace metal is acceptable). The paste is placed on the screen and pressed through, the strained material is then re-collected, put back on top, and strained through again. At this point water can also be added to help thin the paste to the desired consistency, and once this is complete, the WSP is finally ready for use!

Don't be disheartened by the many steps and materials needed for WSP production. The process speeds up with repetition, and the end results are worth the effort. Wheat starch paste has seen continued use for millennia due to its properties, but it remains a sustainable, inexpensive, reversible, and highly stable adhesive in contemporary conservation practice.

***Technical Data***

*Table 2.6* Composition information for several starch pastes. Data collated from Horie [12]

| | *Proportion* | | *Degree of polymerisation ($DP_n$)* | | *Gel formation* |
|---|---|---|---|---|---|
| | *Amylose* | *Amylopectin* | *Amylose* | *Amylopectin* | |
| Wheat | 20–30% | 70–80% | 1300 | 7500 | strong |
| Rice | 16–19% | 81–84% | 1100 | 10,000 | weak |

## SEAWEED GUMS

Consultant: Kyoko Kusunoki

As with the cellulose ethers elsewhere described, some seaweed-derived gums have found widespread use as adhesives and others as gels. While the latter property will be discussed in later sections, most notably for agar-agar, this section will describe the preparation of seaweed gums for use as adhesives or consolidants.

***Material Information***

Precision Required: Medium

*Overview*

Although sometimes lumped in with starches due to their shared polysaccharide backbone, seaweed gums are chemically distinct, derived from galactose monomers rather than glucose. Of the two discussed here, only **fu***nori* (not to be confused with **furu***nori*) is a true seaweed gum, while sodium alginate arrowroot paste is a mixture of seaweed gum and starch. While this technically straddles the line between this book's taxonomies, as its preparation and use more closely aligns with *funori*, the two have been grouped together here.

Perhaps the most widely used seaweed gum, agar-agar, because it is not typically used as an adhesive, will be discussed in the chapter on gels.

*Funori* (布海苔) is a stable water-soluble adhesive derived from several species of the red algal genus *Gloiopeltis*. The seaweed extract is typically referred to as *funoran* in scientific literature, while *funori* is the preferred name for the adhesive in conservation circles [44, 45]. *Funori* can be supplied as unbleached pink-purple or bleached pale-yellow strands of seaweed, or as a dehydrated and purified solid (Jun- or Tri-*Funori*) [46]. As one might expect, the colour of the seaweed will affect the colour of the end solution, with darker-coloured seaweeds yielding a darker solution than the bleached ones. The purified samples are transparent and easier to use, but are also considerably more expensive [46].

Alginate is another polysaccharide, this time derived from numerous 'brown algae' species within the class *Phaeophyceae* and typically used as a gelling agent in medicine. Here it is blended with arrowroot starch to produce a 'dry' mild tack adhesive paste.

*Composition*

***Funori***: The exact composition varies based on the species of seaweed, but *funori* mostly consists of a heterogeneous galactose sulphated polysaccharide (β-d-galactose-6-sulphate–3,6-anhydro-α-l-galactose diad). It is strongly suspected that the sulphurous residues of this structure contribute to prepared *funori*'s characteristic liquidity at room temperature [12, 44, 47]. Although workably consistent, one can expect some variation as factors such as the exact species of *funori* seaweed used, extraction temperature, and the harvesting location can all also affect molecular weight [47].

**Alginate:** Like *funori*, alginate is extracted from seaweed, in this case the 'brown algae' *Phaeophyceae* class. However, in this instance, the alginate must be isolated with treatments of aqueous alkali solutions, most often sodium hydroxide, and precipitated with sodium or calcium chloride. This produces an alginate salt which is then further purified and refined until a water-soluble sodium alginate power is produced. When looking at the seaweed on a dry weight basis, the alginate contents range roughly between 22 and 44%. Alginate is composed from a whole family of aliphatic copolymers that contain blocks of (1,4)-linked β-D-mannuronate and α-L-guluronate residues.

***Preparation:* Funori**

*Tools and Techniques*

**Water**: Research presented by Noriko Hayakawa, Kyoko Kusunoki, and their co-authors (2015) found that water hardness can significantly affect the viscosity of *funori* solutions [48]. Generally, *funori* prepared with hard water is more viscous than that prepared in deionised water. This is even more apparent at higher temperatures with hard water samples measured at nearly twice the viscosity of those prepared in deionised water (approximately 4000 mPa·sto 2000 mPa·srespectively) [48]. It is therefore recommended that all *funori* preparation, particularly the latter stages, be carried out with distilled or deionised water.

**Pre-soaking**: Dry *funori* seaweed can be laden with salts and other undesirable materials and therefore requires both rinsing and soaking prior to 'cooking' or extraction. The general process is to rinse the pieces, soak them for several hours or overnight, and then rinse again before extraction. Most, if not all, should be done with deionised water, but in particular the long soak, final rinse, and extraction. This should remove any final remnants of salt or other impurities and lower the viscosity of the resultant adhesive. Finally, please note that soak times greater than 8 hours have been shown to have the added benefit of reducing the cook time for the extraction [45].

**Filtration**: If using raw seaweed pieces (either bleached or non-), the water-soluble polysaccharide adhesive must be separated from the solid structure of the seaweed. There are two main approaches for this: the tried-and-true cloth filtration and the slightly more sophisticated vacuum filtration. While both will produce a workable consolidant, broadly speaking, the coarser the method, the more plant matter left in solution and the more turbid the end result. While cloth filtration remains the most commonly cited method, surfaces which require a high level of colour precision may benefit from the extra step of suction filtration [45, 46].

**Extraction temperature**: The *funori* adhesive can be extracted from the seaweed over a range of relatively low temperatures, but as ever, the extraction temperature can affect mechanical and chemical properties. In a comparison of *funori* samples extracted at 95°C and 25°C, Hayakawa et al. found that the higher temperature (95 °C) solutions were more viscous and had a higher molecular weight than their 'cool' (25°C) counterparts. However despite their lower viscosity, dried films of the 25°C extractions exhibited higher peel strength than the 95°C samples [47]. Broadly speaking, the warmer the extraction, the stronger and more viscous the solution, but after 90°C any benefit of higher temperatures is offset by the increased risk of burning [45]. Therefore, weak, low-viscosity *funori* can be obtained by low-temperature extractions (20–30 °C) and more viscous solutions are best extracted between 80 °C and 90 °C [46, 49].

*Materials*

Beaker
Hot plate
Balance
Glass stirring rod
Filtration method: very clean fine fabric or nylon strainer; or vacuum filter funnel, Whatman No. 1 filter paper, and Büchner flask beaker, vacuum pump, and tubing.

*Instructions*

Adapted from Swider and Smith 2005, and Gattuso et al. 2020 [45, 46]
Ratio: 3 g *funori*: 100 mL water

1. Using the balance, weigh out the desired amount of *funori* and cut into small (1–4 cm) pieces. Place the pieces into the beaker.
2. Place the beaker under a dripping tap and allow water to drip onto the *funori* for roughly 15–30 minutes. Drain the *funori* after this first rinse, refill the beaker with deionised water, and set aside for several hours (ideally overnight).
3. The next day, tip out the soak water and rinse the seaweed one more time in deionised water. Add the pieces back to the beaker and top up with the desired final volume of water.
4. Optional step: 10 mL ethanol can be added before heating the mixture. This can stabilise the solution, slowing any biological growth, and can lower the surface tension of the mixture, allowing for easier filtration. It's worth noting that the addition of too much alcohol can cause the *funori* to precipitate out of solution.
5. After soaking, heat the mixture to 70–80 °C, stirring occasionally. A small amount of simmering can be tolerated but avoid boiling the mixture. Heat until the seaweed is mostly or completely dissolved. Depending on the specific parts of seaweed used, there might be some sediment or insoluble components.
6. Filtration: if slight colouration of the solution can be tolerated in the treatment, allow the extraction to cool and then strain it through silk, four layers of cheesecloth, or a fine mesh strainer. Retain the liquid and discard the solids. If using fabric offcuts, ensure that these are very clean and ideally free from sizing, bleach, starch, or other additives.
   *Or*
   Assemble the suction filtration, placing the funnel in the Büchner flask and attach your vacuum source. Place the filter paper in the funnel and lightly dampen with deionized water. Turn on the pump to pull the rinse

water through the paper, removing any dust or surface impurities. Discard this water from the funnel and reassemble. Add the solution to the funnel and turn on the pump. Once water stops readily leaching through the paper, turn off the pump. Retain the liquid and discard the solids.

7. Once strained, *funori* can be thinned to the desired consistency (typically less than 1% for consolidation, up to 5% for adhesion) and used straightaway or redried for later use.

*Dehydrating and Rehydrating*

Once filtered *funori* can be dried and quickly rehydrated prior to use [45, 50]. Ideally, the *funori* should be dried in a room with minimal dust (and therefore minimal contamination), but anywhere clean will do.

1. Using a small dropper (or simply pouring from the beaker) cast the *funori* onto some silicone-coated polyester sheets or glass. Please note that *funori* has been found to stick to uncoated Melinex/Mylar and will soak into silicone release paper.
2. Allow the pools to air dry naturally. Once dried, peel or scrape the *funori* from the casting surface and place in a sealable jar of bag. Store in a cool dry place (several months) until needed.
3. To rehydrate, simply soak the pieces in room temperature deionized water until they dissolve. Thin to the desired consistency.

*Storage*

Once dried, the purified *funori* can be kept in a sealed container (jar or food storage box) at room temperature for up to four years. Hydrated samples can be kept refrigerated until the adhesive begins to mould, but given the risk of biological growth, it is typically preferable to rehydrate small amounts immediately prior to use [45].

***Preparation: Alginate – Arrowroot***

Arrowroot starch is derived from the arrowroot plant (*Maranta arundinacea*). The combination of the two materials is reported to produce a paste that is drier than other starch adhesive but is comparable in strength and aging properties. For a more in-depth discussion of starch as a class of materials, see the section on starch pastes. When compared to other starches used in conservation, arrowroot starch has a relatively high percent composition of amylose (18–36%) to amylopectin [51, 52].

*Materials*

Beaker
Hot plate

Balance
Graduated cylinder
Glass stirring rod

*Instructions*

Ratio: 1 g sodium alginate: 100 mL distilled water: 5 g arrowroot

1. Fill a beaker with the desired amount of water and sprinkle the sodium alginate powder on top.
2. Let dissolve for several hours, until no solids are visible.
3. Once the alginate is dissolved into solution, add the dry arrowroot starch powder into the bottom of a separate beaker.
4. Add a small amount (roughly 1/4) of the sodium alginate solution to the arrowroot powder and stir well. Stir in the remaining alginate solution.
5. Place the alginate-arrowroot solution on a hot plate and cook gently until the mixture is clear and free of lumps. The mixture should be steaming lightly but not boiling.
6. Once the mixture is well incorporated and smooth, increase the heat to a low simmer and cook for an additional 15–25 minutes. The resulting paste should be relatively dry and lightly tacky.
7. Make immediately prior to application and use within a day or two.

## GLUE

Consultant: Shayne Rivers

***Material Information***

Precision Required: Low

*Overview*

Unlike its colloquial meaning, which encompasses any and all adhesives, in technical fields 'glue' refers to gelatine-based adhesives derived from animal products. These are mostly made by heating or boiling mammalian skins and bones or the swim bladders, skins, and other connective tissues of fish. Animal glues have, very broadly speaking, somewhat fallen out of favour in conservation, having largely been replaced by synthetic polymers. However, glues are largely reversible, non-toxic, require no solvents, and have remained an important part of the tool kits of paintings conservators, luthiers, furniture makers, and restorers (to name a few) [12, 53–56].

It is important to note that even when compared against material like starch pastes, glue has been in continuous use for an extraordinary duration, with some estimates stretching back 10,000 years [12]. As a result, the presence of animal glue on an object does not necessarily indicate either historic repair or modern conservation activity. Care should be taken before removing unprovenanced glue as even ancient glues can often be reactivated and swollen centuries after their original application.

Animal glues are typically offered in several strengths and formats, both liquid and solid. Due to their long history of use, glues have brought with them inherited systems of description. One such system is that of Bloom strength. The Bloom strength essentially describes how 'squishable' a glue of a given solids composition is, with 'low Bloom' gels requiring less force to compress than 'high Bloom' ones [55].

While Bloom is not always the most helpful descriptor in conservation work, it can be used to evaluate the relative properties of a given animal glue. There is a strong correlation between Bloom strength (also called gel strength), average molecular weight, bond strength, and gel time. This means that the higher the Bloom strength, the higher the molecular weight, the harder the gel, the stronger the **ad**hesive bond, and the lower the working time. Conversely, Bloom strength is negatively correlated with **co**hesion. This means that most medium-to-high Bloom animal glues show poor cohesion and, unless bulked with chalk or other fill material, are generally inappropriate for gap filling [12, 53, 57, 58].

Lastly, Bloom strength is a descriptor typically reserved for mammalian glues. Many fish glues are liquid at room temperature, and therefore cannot be evaluated in terms of their gel strength [12]. They can, however, still be compared in terms of their cohesive strength. As the average molecular weight of fish glues is around 25% that of many animal glues, they generally exhibit lower bond strength (save some exceptions).

*Composition*

The adhesion of animal glues is a property of the gelatine found within them. Gelatine is a water-soluble partially hydrolysed form of collagen, which is an abundant protein found in the connective and supportive tissues of animals. Collagen is composed of a triple helical structure of amino acid chains. When heated, the protein denatures, uncoiling and splitting apart, releasing the long threads which then reorganise into a heterogeneous network. This denaturation temperature is based in numerous factors and varies according to the source of the collagen and the amount of processing it has undergone [12, 59].

Unlike fish-derived glues, mammalian collagen must first be isolated from the other chemical components in the hide, hoof, or other source. This is accomplished by subjecting the animal material to a series of enzymatic, alkaline and/or acidic extractions. The first few processes release the longest chains, which are most loosely bound within the collagen network. As the extractions continue, harsher methods are employed resulting in fragmentation, lower average molecular weights, and diminishing quality [12, 60].

The readiness with which the triple helix unwinds, and the degree of processing therefore required, is directly related to the type and age of the animal. Cross-links between the chains increase as an animal ages, and the collagen networks become more rigid and tightly bound. The more cross-linked the triple helices, the more difficult the protein is to break apart, requiring harsher conditions and higher temperatures. As a result, glues from younger animals, such as rabbits, are typically less heavily processed, and therefore less degraded than those from older animals such as cattle or pigs. Similarly, fish collagen is less likely to cross-link than mammalian collagen and therefore denatures readily at lower temperatures and under less-severe conditions [12, 53, 57].

Collagen derived from fish skins yields a lower MW glue but is less likely to form cross-links and therefore requires minimal processing. Isinglass, a glue traditionally derived from the swim bladder of sturgeon, is nearly pure gelatine. It has a higher molecular weight than other fish glues but, because it is fish derived, it tends to have a lower gelling temperature. This allows for deeper penetration before gelling and good adhesion at lower concentrations. It is also relatively easy to dissolve and yields a high clarity solution [12, 50]. In terms of aging, it is unlikely to cross-link or yellow, but it is more sensitive to elevated RH, becoming brittle with prolonged exposure [57].

*Variability of Materials*

The category of animal glues covers an enormous range of adhesives with very different properties. This variety is rooted in many factors, from the age, species, or body part of the source animal; to the processing methods used, the number of extractions and their quality, and the blending that can occur prior to sale.

Comparing and understanding Bloom strength is the most useful metric in navigating this variability. It is therefore best to source collagen glues from a supplier that provides this number. Generally speaking, Bloom 230–250 is around the right strength for joining wood, while 340–360 is appropriate for gilding [5].

Last, but by no means least, the use of isinglass must be carefully considered, and any sources must be critically evaluated. It's estimated that over 85% of sturgeon species are on the brink of extinction, particularly those from the Ponto-Caspian and Western Pacific regions [61]. However, sustainable sturgeon, both wild and aqua farmed, are available and have been used in conservation [62, 63]. Farmed fish in particular can yield a glue with a yellow cast, but this can be reduced with additional processing [62]. J. Cheney's excellent 2022 article describes working directly with caviar farms to source high-quality fresh swim bladders, and then gives thorough step-by-step guidance on how to trim, dry, and rehydrate those bladders into isinglass [63].

***Preparation: Glue***

Note: due to the nature of the materials, the preparation can have a mildly unpleasant (to some) odour. It should not stink or smell rank, as this likely indicates that the glue has begun to rot. Ensure that you have good ventilation, and maybe don't prepare it right next to the lunchroom.

All mammalian recipes adapted from Rivers and Umney 2013 and Ford 1997 and isinglass recipes from Petukhova and Bonadies 1993 and van Dyke 2009 [50, 54, 55, 64]

*Techniques*

**Warming** – In preparing hide glue, you are not looking to cook it as you would with wheat starch paste but only to warm it. The mixture should not be heated above 65–70°C as this can cause the proteins to break down and will weaken the glue. However, the glue can be kept warm (around 60°C) for a full day's work and then refrigerated until the next day. This cycle of warming and cooling can be done repeatedly, but will slowly break the length of the glue chains, changing the properties. Depending on the cycle of heating and cooling, the glue may need to be replaced after a week or so.

**Ratios** – The provided ratios are a loose starting point rather than a prescriptive suggestion. As discussed in the material overview, the strength of a glue will vary based on the source, processing, Bloom strength, and other factors. There is a great deal of publicly available information on the use of animal glues that can be found on websites from luthiers, furniture makers and restorers, and other woodworkers. However, these recipes can be too strongly adhesive for conservation use, potentially encouraging failure within the substrate rather than cohesively within the adherent.

*Tools*

**Glue Pot** – Woodworking and instrument-making forums are filled with options for warming the glue. These can range from pricey purpose-made glue pots to repurposed slow cookers. In every instance the goal is to warm the glue slowly and evenly. It's helpful to think of melting chocolate, where direct heat can scorch the mixture and ruin the product. If working with glues in any consistent way, it is worth investing in one of these methods. However, at the very small scale that many conservators work with, a *bain marie*/water bath or beaker double boiler on a hot plate will suffice.

**Double boiler** – Select a beaker and a small heat-resistant lidded container so that one nests within the other. The beaker should be large enough to hold the container, leaving enough space for it to be easily withdrawn, but not so much space that it tips over. The lidded container can be as simple as a clean jam jar or specimen container, and the lid does not need to fit too tightly. The mixture will be heated, but never to high enough temperatures to pose a problem for most plastics or commercial glassware. The container should be around two or three times the volume of your desired glue mixture; if you want a final volume of around 10–20 mL glue, a 50 mL container is perfect. When the time comes to warm the glue, fill the larger beaker with just enough water to surround the container without spilling over the rim.

*Method: Solid Mammal Glues*

MATERIALS

Glue warmer, or beaker double boiler and hot plate.
Balance
Graduated cylinder

INSTRUCTIONS

| | | |
|---|---|---|
| Ratios: | 1:6 glue to water (v/v) | hide glue adhesive |
| | 1:15 glue to water (v/v) | Rabbit skin glue canvas size |

1. Decide your warming method.
2. Weigh out the desired amount of glue and add to the lidded container. Measure out the desired amount of water and pour over the glue.
3. Allow the mixture to sit for several hours or ideally overnight. The glue will swell.
4. Seal the lidded container and place within the larger beaker. Add just enough water to surround the container without reaching the rim. Place both on a hot plate.
5. Heat the glue mixture to 55–60 °C, stirring occasionally. At this temperature, the glue should readily dissolve within 10–15 minutes. For the most part, leave the lid on to prevent excessive evaporation.
6. Test the consistency and add water to further thin the mixture if desired. Ideally, when picked up with a brush, the glue should run off in a steady stream. If there are lumps, more water may be required; if it easily breaks into droplets, less water is needed. If the water content needs reducing, simply remove the lid and allow the excess to evaporate, or add some more dry glue [5].
7. Must be used while warm.

*Method: Isinglass*

MATERIALS

Dried swim bladder from reputable, sustainable fishery or art suppliers. See WWF report on sturgeon for regional information [61].
Glue warmer, or beaker double boiler and hot plate
Balance
Graduated cylinder

Nylon strainer or pair of tights
Sheet of silicone release polyester or PET film (such as Melinex or Mylar)
Dropper (pipette or syringe) or wide brush

#### INSTRUCTIONS

Ratio 1:20 g:mL

1. Prepare your warming method.
2. Weigh out the desired amount of solid isinglass (often supplied as dried swim bladder) and add to the lidded container.
3. Fill the small beaker with distilled water so that it completely covers the isinglass and allow it to sit overnight..
4. Drain away the soak water and replace with the desired volume of distilled water.
5. Place the lidded container in the double boiler and warm (do not exceed 55 °C) for an hour. Isinglass will dissolve at lower temperatures than mammalian glues; ensure that your water bath is not too hot.
6. Once the solution is fully dissolved, remove the lidded container from the bath and skim away any scum that might have formed on the surface. Allow mixture to cool slightly and then strain it through a screen or double layer of nylon or silk
7. Allow glue to cool to room temperature and use immediately.
8. Any excess can be redried and stored for later use. See following section, 'Dehydrating and Rehydrating Isinglass'. Once dehydrated, these pieces can be broken up when needed and will dissolve quickly in warm water.
9. Optional. If the dried isinglass has a strong yellow cast, this can be reduced by immersing the samples in methylated alcohol or ethyl acetate and placing on a sonicator/jewellery cleaner for at least 30 minutes. See Hems and Curtis (2015) and for longer discussion [62].

#### DEHYDRATING AND REHYDRATING

Once filtered, isinglass can be dried and quickly rehydrated prior to use. The steps below are adapted from Yana van Dyke's technique (2009) and are also appropriate for the storage of *funori* [50].

1. Ideally, the isinglass should be dried in a room with minimal dust (and therefore minimal contamination), but anywhere clean will do.
2. Using a small dropper (or simply pouring from the beaker), cast the isinglass onto some silicone release paper and allow to air dry naturally. This technique results in fairly uniform pieces which are easy to store until needed.
3. To rehydrate, simply soak the pieces in warm deionized water until they dissolve (some very gentle warming may be required).

**Technical Data**

*Table 2.7* Characteristics and use of a selection of glues. Bloom and mechanical strength are rooted in numerous factors which can yield results across an enormous range of values. Comparative strengths are provided as a loose starting point but are by no means comprehensive [53, 58].

| | *pH* | *Bloom (comparative)* | *Colour (dry film)* | *Solutions* | *Use* | |
|---|---|---|---|---|---|---|
| Bone glue | 5–7 | Low to medium<br>160–180 | Dark red-brown | 25–50% v/v adhesive | Adhesive – wood.<br>Less flexible, more penetrating than other glues. Can produce a strong odour during preparation and use | [53, 55, 65] |
| Hide | 6–7.5 | High<br>330–370 (bovine hide Kremer) | Amber (dark to near transparent) | 10–19% v/v adhesive | Adhesive – Wood, book binding, canvas size<br>Strong and more flexible than bone glue. Wide range in source and extraction. | [53, 55, 65] |
| Rabbit skin | 5–7.5 | High<br>340–360 | Pale amber | 10–19% v/v adhesive<br>7% (w/v) size | Canvas size<br>Adhesive – canvas, book binding<br>Binder – gilding, distemper. | [53, 55, 65] |
| Liquid fish glue | 4–7.5 | - | Amber | - | Adhesive – Wood, ceramic.<br>High initial tack, elastic, and slow setting. | [53, 55, 65] |
| Granular fish glue | 6–7 | 575–625 (Kremer) | Pale yellow | - | | [53, 55, 65] |
| Isinglass | 4–6 | Medium to high | Transparent | 3% (w/w) adhesive<br>1–1.5% (w/v) consolidant | Light fast, low refractive index | [53, 55, 65] |

## References


[1] Hxtal Adhesive LLC, "Hxtal NYL-1," Technical Data Sheet.
[2] Epoxy Technology, "Epo-tek 301 Technical Data Sheet," 2021. [Online]. Available: www.epotek.com
[3] Epoxy Technology, "Epo-tek 301-2 Technical Data Sheet," 2021. [Online]. Available: www.epotek.com
[4] Huntsman, "Araldite - Advanced Materials," Data Sheet, 2020. [Online]. Available: www.huntsman.com
[5] A. van Giffen, "Publication Contribution - Conservator's Cookbook," [Personal Communication] Aug. 03, 2024.
[6] S. P. Koob, "Tips and tricks with epoxy and other casting and molding materials," *Objects Specialty Group Postprints*, vol. 10, pp. 158–172, 2004.
[7] J. L. Down, "The yellowing of epoxy resin adhesives: report on natural dark aging," *Studies in Conservation*, vol. 29, no. 2, pp. 63–76, 1984, doi: 10.1179/sic.1984.29.2.63
[8] Huntsman, "Araldite 2020," Technical Data Sheet, 2012. [Online]. Available: www.araldite2000plus.com
[9] Huntsman, "Araldite 2011," Technical Data Sheet, 2014. [Online]. Available: www.araldite2000plus.com
[10] Huntsman, "Araldite 2012," Technical Data Sheet, 2015. [Online]. Available: www.araldite2000plus.com
[11] E. G. Karayannidou, D. S. Achilias, and I. D. Sideridou, "Cure kinetics of epoxy–amine resins used in the restoration of works of art from glass or ceramic," *European Polymer Journal*, vol. 42, no. 12, p. 3311, 2006, doi: 10.1016/j.eurpolymj.2006.08.025
[12] C. V. Horie, *Materials For Conservation: Organic Consolidants, Adhesives, and Coatings*. Butterworths series in conservation and museology. London; Boston: Butterworths, 1987.
[13] C. Stavroudis and S. Blank, "Solvents & sensibility," *WAAC Newsletter*, vol. 11, no. 1, pp. 2–10, 1989.
[14] A. Vinçotte, E. Beauvoit, N. Boyard, and E. Guilminot, "Effect of solvent on PARALOID® B72 and B44 acrylic resins used as adhesives in conservation," *Heritage Science*, vol. 7, no. 1, 2019, doi: 10.1186/s40494-019-0283-9
[15] S. P. Koob, "Manipulating materials: preparing and using Paraloid B-72 adhesive mixtures," *Objects Specialty Group Postprints*, vol. 25, pp. 1–8, 2018.
[16] O. Chiantore and M. Lazzari, "Characterization of acrylic resins," *International Journal of Polymer Analysis and Characterization*, vol. 2, no. 4, pp. 395–408, 1996, doi: 10.1080/10236669608033358
[17] M. R. Schilling, "The glass transition of materials used in conservation," *Studies in Conservation*, vol. 34, no. 3, pp. 110–116, 1989, doi: 10.1179/sic.1989.34.3.110
[18] CTS SRL, "Aquazol Technical Data Sheet," 2010.
[19] Talas, "Aquazol," Talas. Accessed: Aug. 29, 2023. [Online]. Available: https://www.talasonline.com/Aquazol
[20] J. Arslanoglu, "Aquazol as used in conservation practice," *WAAC Newsletter*, vol. 26, no. 1, 2004. [Online]. Available: https://cool.culturalheritage.org/waac/wn/wn26/wn26-1/wn26-105.pdf
[21] B. Ebert, B. Singer, and N. Grimaldi, "Aquazol as a consolidant for matte paint on Vietnamese paintings," *Journal of the Institute of Conservation*, vol. 35, no. 1, pp. 62–76, 2012, doi: 10.1080/19455224.2012.672813
[22] Eastman, "Butvar polyvinyl resins: properties and uses," 2017. [Online]. Available: www.eastman.com/butvar
[23] H. Shockey, "Structural repair of plaster sculpture using polyvinyl butyral resin," *Objects Specialty Group Postprints*, vol. 24, pp. 136–146, 2017.

[24] Synthomer, "M30 Technical Data Sheet," 2023.
[25] Synthomer, "M40 Technical Data Sheet," 2023.
[26] Synthomer, "M50 Technical Data Sheet," 2023.
[27] G. Singh, "AW: [External]Contact form from the Mowital webstie," Jan. 27, 2023.
[28] Dow Chemical Company, "PARALOID™ B-44 Solid Grade Thermoplastic Acrylic Resin." [Online]. Available: http://www.dow.com
[29] Dow Chemical Company, "PARALOID™ B-48N Solid Grade Thermoplastic Acrylic Resin Provides Unique Adhesion, Toughness, Flexibility, and Outdoor Durability." [Online]. Available: http://www.dow.com
[30] Dow Chemical Company, "PARALOID™ B-67 Solid Grade Thermoplastic Acrylic Resin Hydrophobic Polymer Providing Excellent Water Resistance." [Online]. Available: http://www.dow.com
[31] Dow Chemical Company, "PARALOID™ B-72 Solid Grade Thermoplastic Acrylic Resin General Purpose Resin Capable of Forming Softer Films." [Online]. Available: www.dow.com
[32] Dow Chemical Company, "PARALOID™ F-10 Resin Thermoplastic Solution Resin for Flexible Laminate Coating Applications Over Oleoresinous and Other Solvent Sensitive Finishes." [Online]. Available: http://www.dow.com/coating
[33] R. L. Feller and M. Wilt, Evaluation of Cellulose Ethers for Conservation, in *Research in conservation*. Los Angeles: The Getty Conservation Institute, 1990.
[34] Ashland, "Klucel hydroxypropylcellulose: physical and chemical properties." [Online]. Available: https://www.ashland.com/file_source/Ashland/Product/Documents/Pharmaceutical/PC_11229_Klucel_HPC.pdf
[35] Talas, "Conservation, archival & bookbinding supplies," Talas. Accessed Aug. 29, 2023. [Online]. Available: https://www.talasonline.com/
[36] R. Belard, H. Higuchi, and J. Perry, "Furunori (aged wheat starch paste): challenges of production in non-traditional settings," *Journal of the Institute of Conservation*, vol. 32, no. 1, pp. 31–51, 2009, doi: 10.1080/19455220802630735
[37] N. Hayakawa *et al.*, "Characterization of furunori (aged paste) and preparation of a polysaccharide similar to furunori," *Studies in Conservation*, vol. 52, no. 3, pp. 221–232, 2007, doi: 10.1179/sic.2007.52.3.221
[38] M. Matsumaru, "Wheat starch paste: a study of cooking profiles and adhesive properties across preparation recipes," *Journal of the Institute of Conservation*, vol. 44, no. 1, pp. 25–46, 2021, doi: 10.1080/19455224.2020.1864747
[39] American Institute for Conservation (AIC), "BPG adhesive recipes and tips," Wiki. Accessed: Aug. 29, 2023. [Online]. Available: https://www.conservation-wiki.com/wiki/BPG_Adhesive_Recipes_and_Tips
[40] B. F. Miller and W. Root, "Long-term storage of wheat starch paste," 1991.
[41] Making Wheat Starch Paste, (Nov. 19, 2018). Accessed: Aug. 30, 2023. [Online Video]. Available: https://www.youtube.com/watch?v=cm9yd35K4yQ
[42] M. Matsumaru, "Paste-making tips with two recipes," The Book & Paper Gathering. Accessed: Aug. 30, 2023. [Online]. Available: https://thebookandpapergathering.org/2016/10/20/paste-making-tips-with-two-recipes/
[43] M. Matsumaru, "Japanese horsehair sieves: who makes them and how to use them?," The Book & Paper Gathering. Accessed: Oct. 05, 2023. [Online]. Available: https://thebookandpapergathering.org/2022/01/20/japanese-horsehair-sieves-who-makes-them-and-how-to-use-them/
[44] R. Tuvikene *et al.*, "Funorans from Gloiopeltis species. Part I. extraction and structural characteristics," *Food Hydrocolloids*, vol. 43, pp. 481–492, 2015, doi: 10.1016/j.foodhyd.2014.07.010
[45] J. R. Swider and M. Smith, "*Funori*: overview of a 300-year-old consolidant," *Journal of the American Institute for Conservation*, vol. 44, no. 2, pp. 117–126, 2005, doi: 10.1179/019713605806082329

[46] C. Gattuso, L. Campanella, C. V. L. Giosafatto, L. Mariniello, and V. Roviello, "The consolidating and adhesive properties of *funori*: microscopy findings on common and ancient paper samples," *Journal of Cultural Heritage*, vol. 48, pp. 153–160, 2021, doi: 10.1016/j.culher.2020.11.013

[47] N. Hayakawa, K. Kida, T. Ohmura, N. Yamamoto, K. Kusunoki, and W. Kawanobe, "Characterisation of *funori* as a conservation material: Influence of seaweed species and extraction temperature," *Studies in Conservation*, vol. 59, pp. S230–S231, 2014, doi: 10.1179/204705814X13975704320035

[48] N. Hayakawa, T. Omura, Y. Hara, K. Kusunoki, K. Kida, and T. Honda, "Effect of extraction conditions on the physical properties of *funori* extracts: Temperature, seed, and water hardness," in *フノリ抽出物の物性に及ぼす抽出条件の影響—温度・種・水の硬度—(早川典子、大村卓也、原由宇稀、楠京子、貴田啓子、本多貴之) 文化財保存修復学会第37回大会 京都工芸繊維大学*, Kyoto, Japan, Jun. 2015.

[49] T. Geiger and F. Michel, "Studies on the polysaccharide Jun*Funori* used to consolidate matt paint," *Studies in Conservation*, vol. 50, no. 3, pp. 193–204, 2005, doi: 10.1179/sic.2005.50.3.193

[50] Y. van Dyke, "Sacred leaves: the conservation and exhibition of early Buddhist manuscripts on palm leaves," *The Book and Paper Group Annual*, vol. 28, 2009, [Online]. Available: https://cool.culturalheritage.org/coolaic/sg/bpg/annual/v28/bpga28-17.pdf

[51] L. B. O. Rodrigues, C. M. Veloso, R. C. F. Bonomo, L. B. Rodrigues, L. A. Minim, and R. A. S. Costa, "Rheological and textural studies of arrowroot (Maranta arundinacea) starch–sucrose–whey protein concentrate gels," *Journal of Food Science and Technology*, vol. 55, no. 8, pp. 2974–2984, 2018, doi: 10.1007/s13197-018-3215-x

[52] G. F. Nogueira, F. M. Fakhouri, and R. A. De Oliveira, "Extraction and characterization of arrowroot (Maranta arundinaceae L.) starch and its application in edible films," *Carbohydrate Polymers*, vol. 186, pp. 64–72, 2018, doi: 10.1016/j.carbpol.2018.01.024

[53] N. C. Schellmann, "Animal glues: a review of their key properties relevant to conservation," *Studies in Conservation*, vol. 52, no. sup1, pp. 55–66, 2007, doi: 10.1179/sic.2007.52.supplement-1.55.

[54] F. Ford, "I'll stick with hide glue…," Frets.com. Accessed: Aug. 31, 2023. [Online]. Available: http://www.frets.com/FretsPages/Luthier/Data/Materials/hideglue.html

[55] S. Rivers and N. Umney, *Conservation of Furniture*. Abingdon: Routledge, 2013.

[56] W. P. Edwards, "Choosing and mixing hot hide glue," FineWoodworking. Accessed: Aug. 31, 2023. [Online]. Available: https://www.finewoodworking.com/2020/03/20/choosing-and-mixing-hot-hide-glue

[57] A. Bridarolli, A. A. Freeman, N. Fujisawa, and M. Łukomski, "Mechanical properties of mammalian and fish glues over range of temperature and humidity," *Journal of Cultural Heritage*, vol. 53, pp. 226–235, 2022, doi: 10.1016/j.culher.2021.12.005

[58] A. Melià-Angulo, L. Fuster-López, and A. V. Escuder, "Study of the mechanical properties of selected animal glues and their implication when designing conservation strategies," *CeROArt. Conservation, exposition, Restauration d'Objets d'Art*, no. HS, Art. no. HS, May 2017, doi: 10.4000/ceroart.5152

[59] J. Hubbard, "Animal Glues," in *Handbook of Adhesives*, New York: Van Nostrand Reinhold Co., 1977. Accessed: Aug. 31, 2023. [Online]. Available: http://archive.org/details/handbookofadhesi00skei

[60] G. Ntasi *et al.*, "Proteomic characterization of collagen-based animal glues for restoration," *Journal of Proteome Research*, 2022, doi: 10.1021/acs.jproteome.2c00232

[51] R. Reinartz and P. Slavcheva, "Saving sturgeons: A global report on their status and suggested conservation strategy," World Wildlife Federation (WWF), 2016. [Online]. Available: https://wwfint.awsassets.panda.org/downloads/Saving_sturgeons_A_global_report_on_their_status_and_suggested_conservation_strategy.pdf

[62] E. S. Hems and A. E. Curtis, "Decolourising isinglass derived from aqua-farmed sturgeon by sonication," *Journal of the Institute of Conservation*, vol. 38, no. 2, pp. 188–199, 2015, doi: 10.1080/19455224.2015.1068199

[63] J. E. Cheney, "Extracting isinglass from fresh air bladders of sustainable Canadian-sourced wild Atlantic sturgeon," *Journal of the Canadian Association for Conservation*, vol. 46, 2022, Accessed: Jun. 12, 2024. [Online]. Available: https://www.cac-accr.ca/publications/extracting-isinglass-from-fresh-air-bladders-of-sustainable-canadian-sourced-wild-atlantic-sturgeon/

[64] T. Petukhova and S. D. Bonadies, "Sturgeon glue for painting consolidation in Russia," *Journal of the American Institute for Conservation*, vol. 32, no. 1, pp. 23–31, 1993, doi: 10.1179/019713693806066483

[65] Kremer Pigmente, "63000 – 63080 technical data (glues)," Technical Data Sheet. Accessed: Apr. 29, 2023. [Online]. Available: https://www.kremer-pigmente.com/elements/resources/products/files/63000-63080e.pdf

# 3 Clays and Gels

**Caution**

Either through intentional cleaning process or unintentional residues left behind, all the materials in this section have the potential to permanently alter the character of an object's surface.

Undesirable materials left by gels or clay-mediated cleaning systems have been the subject of considerable research, but firm conclusions remain elusive. To what extent these residues are left behind and how damaging such compounds may be are both contingent on the specific combination of materials used and the chemical character and physical topography or porosity of the surface being treated.

Each subsection will indicate where these are known issues, but the topic as a whole is the subject of evolving research and should be carefully evaluated during the risk-benefit planning that precedes any conservation treatment.

## CLAYS

Consultant: Jennifer Dinsmore

### *Material Information*

Precision Required: Low

*Overview*

Natural clays are composed of chemically inert silicates which can support the use of a wide range of solvents and complex cleaning solutions. They have seen particular use as a poultice material for architectural and sculpture conservation where they have been employed to clean and/or desalinate stone and masonry [1]. There is also some description of use within paper conservation, but this seems to have fallen due to concerns about residues and the rise of gelled systems.

DOI: 10.4324/9781003391708-4

Clays are most frequently used as poultices with one of two aims. They are either employed to hold a solvent or cleaning mixture in close contact with a surface or to provide a receptacle for undesirable materials which have been wicked out of the substrate. For example, kaolin has been used as a carrier for acetone in the removal of aged adhesive tape from paper surfaces while attapulgite is used in desalination treatments for stone [2].

Many different types of absorbent inorganic materials have been used in conservation, but current literature and supplier catalogues indicate that attapulgite (palygorskite), sepiolite, kaolin, and Laponite RD are the compounds currently most commonly employed.

### *Composition*

Attapulgite (also known as palygorskite) and sepiolite are primarily hydrated magnesium silicates, and both can absorb a large volume of water due to their open structure. These minerals are 'fibrous' and organize as a stack of long straws, thus allowing capillary action and evaporation to drive the water up and through, from one end to the other. Although typically discussed in terms of water uptake, these clays will also absorb polar and nonpolar solvents and many oils [3–5].

Kaolin is a white or grey naturally occurring mineral, consisting mostly of kaolinite, a hydrated aluminium silicate [6]. Long recognised as a key component of porcelain, kaolin has come into common use in a wide range of industrial, medical, and cultural applications [3]. It is soft and smooth to the touch, and like other clays it is inert and can absorb both polar and non-polar solvents. Unlike attapulgite and sepiolite, kaolin is non-fibrous, instead consisting of layers of hexagonal arrays of silicon tetrahedra and octahedral aluminium hydroxides stacked into large consecutive sheets, a structure known as a kaolinite book [7].

Laponite RD is a synthetic nano-clay that, while not completely inert, has the benefit of visual transparency and has proven stable in solvent systems with pH of 3–10 [8]. Laponite RD [ layers of disk-like octahedral Mg and Li sheets sandwiched between layers of tetrahedral silica [9]. This structure draws it into closer analogy to kaolin than attapulgite or sepiolite, but its working properties are closer to gelling systems than to other clays. In fact, Laponite is technically more of a Pickering emulsifier (see '(Micro)Emulsions') than a traditional clay, but as its preparation and use align it with this broad grouping, it has been included here.

### ***Preparation: Natural Clays***

#### *Techniques and Tips*

**Residues**: Attapulgite and sepiolite are both known to, in some circumstances, leave behind a thin grey layer. Although clay residues are inert and not necessarily physically damaging in themselves, they can change the appearance and prove very difficult to remove [10].

**Application**: As discussed above, all clays have the potential to leave behind residues, but how much of a problem this is will depend on the substrate. In desalination treatments or cleaning protocols for stone and masonry in which full saturation of the object is acceptable or desirable, the object can be wetted, wet poultices applied, and covered with cling film to slow evaporation. Once whatever cleaning solution is given time to work, the cling film is removed and the clay poultice is allowed to partially dry in situ. This allows any cleaning materials or undesirable straining or salts to be pulled from the stone surface and up into the poultice [11]. Although less potentially damaging than Laponite, clay poultices should not be allowed to fully dry on the surface. Any soiling which had been pulled from the substrate would simply be deposited on the surface, thus defeating the point of cleaning.

The risk of residue deposition can be somewhat reduced through the use of a barrier layer of Japanese tissue or spun polyester (Hollytex). This also simplifies poultice removal, but can lessen the close conformation and surface contact desired for best cleaning outcomes.

**Mixing**: Of the preparations in this book, clay poulticing is perhaps the most flexible and subjective. While some precision is required for any complex cleaning systems to be loaded into the clay, the ratio of clay to aqueous solution depends entirely on the preferences of the conservator and the needs of the project. Clay which needs to closely conform to an uneven vertical surface will need to be wetter than a clay poultice that will sit on top of a horizontal surface, but not so wet it will slide down the face. As a result, there are no specific ratios or parts to follow for this preparation.

*Materials*

Volumetric measure

Stirrer: glass, wood, or metal spoon or spatula is acceptable. Unless using metal-reactive compounds (such as chelation agents) in the solution, the choice of material should not affect the mixture.

Mixing container: anything from a beaker up to a large bucket

*Instructions*

1. If using an aqueous system, prepare your cleaning solution as per the instructions discussed in Chapter 4.
2. Add clay powder to mixing container and then slowly add aqueous system or straight solvent until the desired consistency is met.

3 Stir well and adjust pH if required. Rember that some clays carry an independent pH which will need to be evaluated alongside the other poulticing components.

***Preparation: Laponite RD***

*Techniques and Tips*

**Residues**: Small amounts of Laponite RD has been found impacted in the micro-topography of ceramic objects which were previously treated with the gels to remove impacted staining [12]. The amount of residue appears to positively correlate to the percent concentration of the gel, with higher concentrations resulting in higher levels of deposition [12]. However, surface topography was also seen to be a significant factor, with Laponite crystals also becoming wedges or trapped within fine pores and microcracks. Finally, although inconclusive, it is perhaps worth mentioning that although no damage was observed, there were indications that the gel can provoke the movement of cations within the glaze layer [12]. As with the natural clays, the presence of residues can be lessened through use of a barrier layer. Gampi paper in particular has been shown to limit (if not outright prevent) the deposition of Laponite on a paper substrate [13].

**Solvents**: Laponite can be made with up to 1/3 of liquid volume of solvent, but the solvent must be miscible with water, and large quantities of non-polar solvents, such as xylene, won't work.

**Removal**: The Laponite mixture should be both applied and removed while wet. Laponite should not be allowed to dry onto surfaces where it can become strongly adhered and difficult to remove.

*Method: Aqueous Laponite Mixture*

MATERIALS

Weighing boats
Balance
Graduated cylinder
Magnetic stirrer is ideal, but stirring rod (of any material) will be fine

CONTAINER WITH LID INSTRUCTIONS

1. Weigh out the desired amount of Laponite powder (most literature records 2–5% w:v).
2. Measure out the full volume of water and pour into lidded container.
3. Stir water at medium speed and slowly add the Laponite powder until well mixed.

4. You can either keep stirring for 30 minutes or close the container and allow the mixture to sit and thicken for several hours.
5. After rest, lightly stir and if necessary, check and adjust the pH.

*Method: Solvent Laponite Mixture*

MATERIALS

Weighing boats
Balance
Graduated cylinder
Magnetic stirrer is ideal, but stirring rod (of any material) will be fine
Container with lid

INSTRUCTIONS

1. Weigh out the desired amount of Laponite powder (most literature records 2–5% w:v).
2. If preparing a 2:1 water to solvent gel, the aqueous portion should measure roughly 2/3rds of the desired final volume. Measure out water and pour into lidded container.
3. Stir water at medium speed and slowly add the Laponite powder until well mixed.
4. Allow the mixture to rest for approximately 20 minutes.
5. Add in the desired solvent up to roughly 1/3 of the desired final volume.
6. Stir well and allow to rest for several hours.
7. After rest, lightly stir and if necessary, check and adjust the pH.

***Technical Data***

*Table 3.1* Selection of clays used in conservation with CAS number and pH information

| *Name* | *CAS number* | *pH* | *Source* |
|---|---|---|---|
| Attapulgite/Palygorskite | 12174-11-7 | 7.5–9 | [14] |
| Sepiolite | 63800-37-3 | 8.3 | [15] |
| Kaolin | 1332-58-7 | 4.0–5.0 | [16] |
| Laponite | 53320-86-8 | 9.8 (2% suspension) | [17] |

## SPREADABLE GELS

Consultant: Joyce Townsend

***Material Information***

Precision Required: Low

*Overview*

Continuing with the theme of preparation and use-based taxonomies, spreadable gels are thus grouped for their shared ability to be smoothly spread over a surface. This behaviour and certain elements of their preparation are often the only shared characteristics between these chemically and physically diverse systems. Confusing matters further, several of the materials listed below are identified as surfactants within the industrial literature but are referred to as gelling agents or even adhesives within conservation circles, while others are also used as adhesives. As a result, some of the materials in this section are also discussed elsewhere in the text and where relevant this will be indicated.

Spreadable, or thixotropic, gels are primarily used as cleaning systems as they give excellent conformation to surface topography and allow for very controlled application of solvents or surfactant mixtures. However, as with all the techniques in this section, there is a strong possibility of leaving excess material behind, and the suitability of the surface must be carefully considered prior to treatment. It was found that Carbopol could be completely cleared from smooth, uniform, 'ideal' surfaces, but also that gel particles can become trapped within microcracks of porous and uneven surfaces, where they prove extremely difficult to clear [18]. Conservators are strongly advised to see *Solvent Gels for the Cleaning of Works of Art: The Residue Question* (2004) for a more thorough discussion [19]. An open access pdf of the entire publication is available through The Getty's website [20].

The materials in this section include: Carbopol, Pemulen, methyl cellulose (MC), and xanthan gum (XG).

Please note: The gels should never be allowed to dry onto the surface of the object as they can form a tightly adhered layer which can be extremely challenging to remove.

*Composition and Characteristics*

Due to the variety and complexity outlined previously, chemical composition is only very briefly summarised.

CARBOMER/CARBOPOL

Carbomer is the generic name for a series of polyacrylic acid polymers more widely known by the trademark Carbopol. Carbopol is supplied as a slightly acidic powder and is poorly wetted by water unless the pH is raised above 6. In the presence of a small amount of alkali and water, these acidic chains will uncoil, cross-link, and form a stable network of very high molecular weight [21]. Chemically, the addition of a neutralising base ionises the polymer, generating negative charges along the backbone. The like-charged anions then repulse one another, causing the molecule to very rapidly uncoil into an extended structure [22].

This neutralisation can be accomplished through the addition of any number of alkaline compounds, but Ethomeen and Armeen (proprietary amine ethoxylates currently produced by Nouryon) are by far the most widely represented in this context.

Several carbomer polymers are available, but as Carbopol 980, 934, 940, and 941 are the formulations that feature most frequently in the conservation literature only these will be discussed in detail [4, 21, 23].

### PEMULEN

Like Carbopol, Pemulen is one of a class of acrylate/C10-30 alkyl acrylate copolymers. As such, it consists of a collection of polyacrylic acid backbones that are cross-linked with long-chained methacrylates [24]. The polymer has the amphiphilic character of a surfactant, with both a lipophilic methacrylate group and a hydrophilic acrylic acid. This allows it to act as an emulsifier, facilitating oil into water (O/W) emulsions without the addition of other surfactants [24]. See the sections 'Surfactants' and '(Micro)emulsions' for more detail on these mixtures.

Pemulen is available in two varieties: TR1 and TR2. When within its optimum pH range of 4–5.5, TR1 will emulsify up to 30% of the dispersed (oil) phase of an O/W emulsion, but outside that range (3–4 and 5.5–11) this amount is closer to 20% [21, 25]. At the same pH optima, TR2 can emulsify more oil than TR1 (60–80% by weight), but both these values are dependent on the specific solvents being used [25]. A little Pemulen goes a long way, and TR2 is recorded as forming stable O/W emulsions at very low concentrations of 0.5–1%(w/v) [24].

Although very much an emulsifier rather than a gelling agent, Pemulen has particular strength when compared against Carbopol for its ability to suspend nonpolar soiling within an aqueous gel system. It therefore performs the function of surfactant without the addition of further reagents. This also means that once the gel is cleared, no additional steps are required to remove excess surfactants [26].

### METHYL CELLULOSE

Cellulose ethers are a class of naturally derived or synthetic polysaccharides which are distinguished by their different conformation architecture, substitutions, and branch character. As these materials are widely used as adhesives, their structure and character are discussed in greater detail in Cellulose Ethers Section in Chapter 2. As one specific cellulose ether, methyl cellulose (MC), is frequently cited as a gelling agent and as the preparation of the gel varies *slightly* from the adhesive, mostly in terms of additives, discussion of its formulation is included here.

### XANTHAN GUM

Xanthan gum (XG) is another naturally occurring polysaccharide, but this one is exuded from *Xanthomonas campestris* bacteria as a by-product of fermentation [21]. Xanthan gum has a very high molecular weight, and while the exact configuration of the molecule is dependent on the pH of the solution, it will form stable gels at any pH between 2 and 12 [21, 27, 28]. XG gels exhibit an inconsistent structure which allows for the formation of pockets that suspend oil and non-water-miscible solvents within the gel network [21]. This means that xanthan gum, like other rheology modifiers in this section, acts as a gelling agent and emulsifier with 2% xanthan gel holding up to 20% w/w of non-miscible solvent [21].

Water-miscible solvents such as acetone and alcohols can also be dissolved directly into xanthan gels, up to 40–50% total volume, while non-miscible solvents can be emulsified in small hydrophobic 'pockets' that form within the structure. Xanthan gels exhibit sheer thinning, where agitation, such as that of a small brush, will break apart those pockets releasing the non-miscible liquids [21, 26]. When compared against other rheology modifiers, xanthan gum is straightforward to prepare, easy to clear with white spirit or water, inexpensive, and sustainable [26, 29]. Unique to the rheology modifiers discussed in this section, xanthan gums can also be used as a modifying additive, changing the characteristics of non-spreadable/rigid gels, a technique which is discussed in greater detail in Cellulose Ethers Section in Chapter 2.

***Preparation: Carbopol Gel***

*Techniques and Tips*

**Caution**: Can leave residues on uneven or porous surfaces. Additional reagents and solvents may require PPE precautions. Do not allow gels to dry onto the surface of the object.

**Lead time**: 1–2 days

**Useful concentrations**: 0.5–3 (wt. %) [23]

**Formulation selection**: Carbopol and Ethomeen are both available under a variety of different formulations, and the properties of both can vary significantly. Carbopol 980 is often chosen as a good starting point (and is listed in the recipes below) because it is a simply *enormous* molecule, with a molecular weight recorded between 4 million and 4.5 *billion*, which is thought to make it easier to clear [4, 30].

The amine formulation used, be it Ethomeen or Armeen, is also an important consideration as each exhibits different solvent compatibilities and is therefore better suited to the preparation of one type of gel over another. For example, Ethomeen C25 is best used in the preparation high polarity solvent gels (acetone, etc.), Ethomeen C12 is used for mid-polarity solvents, and Armeen 2 °C is best for non-polar solvent gels (xylene, petroleum spirts, etc.) [4, 26]. Again, however, although Carbopol is most often used in solvent gels, it can be neutralised with other compounds to form aqueous systems. See Table 3.4 in the technical data for this Section for additional information.

Finally, in correspondence with Lubrizol, the technical advisor stated that some of the formations cited in the conservation literature – namely, 934, 940, and 941 – are polymerized in benzene and will therefore contain small amounts of this solvent. He then suggested 974P NF, 980NF, and 981NF as benzene-free replacements, but as these are currently unattested in the conservation literature, further research is required [31].

**Solvents**: A quick note on solvents and co-solvents. When using more than one solvent, remember to take this into account when planning your measures. However, please note that volume is not conserved. This means that 50 mL of one solvent mixed with 50 mL of another will not always result in a total solution volume of 100 mL. The degree of variation will depend entirely on the volume and the specific solvents used, and the effect may well be negligible, but it is worth reiterating here. The calculations given below will yield approximate mixtures.

For example: I want to make a 1.5% MC solution in a mixture of water and 10% isopropanol (w:v).

1.5 g MC
100 mL total solvent

$$10\% \text{ isopropanol} \times 100 \text{ mL total solvent} = 10 \text{ mL isopropanol}$$
$$100 \text{ total solvent} - 10 \text{ mL isopropanol} = 90 \text{ mL water}$$

**pH**: Carbopol is supplied as an acidic white solid and requires the addition of a neutralising base. Aqueous dispersions of Carbopol have a pH ranging from 2.8 to 3.2, but as the pH increases the mixture swiftly thickens, hitting a maximum viscosity plateau around pH 5 [32]. In order to obtain fully developed gels, the final pH should lie somewhere between 5.5 up to 8.5 [31].

Please note that pH readings should be done with pH indicator papers, rather than a pH meter. The meter will not give correct readings for either Carbopol or Pemulen systems. Although the exact mechanism has not yet been determined, it is thought that the polymer may interact with the electrode affecting all results [33].

**Temperature**: Carbopol gels are largely temperature-stable, tolerating sustained warming up to 65°C, and some formulations can withstand freeze-thaw cycles without changes in viscosity [34]. However, the stability of any non-aqueous solvent additives (such as those used in solvent gels) is not described.

**Disposal**: As always, check all supplier-provided SDS for updated disposal information. However, it is worth stating that Ethomeen C12 is a known aquatic pollutant and should never be tipped down the drain.

*Method: Carbopol Solvent Gel*

Methods adapted from Rivers and Umney [4], Wolbers [35], and Lubrizol [36]

MATERIALS

Ratios for approximate 1.5% w/v solution

1 g Carbopol 980
10 mL amine: Ethomeen C25 (polar solvent gels) or Ethomeen C12 (medium-polar solvent gels), or Armeen 2C (non-polar)
50 mL solvent
7–10 mL water
Weighing boats
Balance
Volumetric measure: graduated cylinder, syringe, etc.
Magnetic stirrer is ideal, but stirring rod (of any material) is acceptable.
Container with well-sealing lid
Small beaker
Pipette or dropper
pH indicator papers

INSTRUCTIONS

1. Using the volumetric measure, measure out the amine and add it to the lidded container.
2. Weigh out Carbopol.
3. Stir the amine mixture until a small vortex forms and dust the Carbopol powder on the surface.
4. Mix well until the Carbopol is dispersed (some small lumps are okay).
5. Using the same volumetric measure (no need to wash it), measure out the desired solvent and add it to the Carbopol-amine mixture.
6. Add any cleaning agents or non-aqueous co-solvents. If using any co-solvents, be sure to deduct this from the total as described in 'Techniques and Tips: Solvents'.
7. Mix until homogenous.
8. If desired, check the pH using indicator papers, and adjust as needed. Mix well.
9. Once the solvent is fully incorporated and the pH value is met, let the mixture rest while you pour a small amount of deionised water into a small beaker.
10. Remove the lid from the gel mixture, and using the pipette or dropper, add a drop or two of water to the gel mixture.
11. Stir well.
12. The gel mixture will continue to thicken as the water is incorporated. Continue to add water dropwise until the desired consistency is reached.
13. Make any final pH measurements and adjustments
14. Seal the lidded container and allow the gel to rest overnight. This will allow the polymer to unfurl and gel more completely. Mix gently prior to use.

*Method: Aqueous Carbopol Gel*

Methods adapted from Rivers and Umney [4], Wolbers [35], and Lubrizol [36]

MATERIALS

Ratio for 1% Carbopol mixture to pH 6–7
1 g Carbopol
98 mL deionised water
2–2.5 mL 18% (w/v) sodium hydroxide NaOH in water (see technical data section for other alkali ratios)
Weighing boats
Balance
Volumetric measure: graduated cylinder, syringe, etc.
Magnetic stirrer is ideal, but stirring rod (of any material) is acceptable.
Container with well-sealing lid
Small beaker
Pipette or dropper
pH indicator papers

INSTRUCTIONS

1. If not already made, prepare the alkaline solution and set aside.
2. Using the volumetric measure, measure out the deionized water and add it to the lidded container.
3. Add any desired surfactants or chelation agents. Keep in mind the changing pH conditions particularly with the later addition of alkali.
4. Weigh out Carbopol
5. Stir the water until a small vortex forms and dust the Carbopol powder on the surface.
6. Mix well until the Carbopol is dispersed (some small lumps are okay).
7. Continue stirring and add in the alkali solution.
8. Mix until homogenous.
9. If desired, check the pH using indicator papers, and adjust as needed. Mix well.
10. Seal the lidded container and allow the gel to rest overnight. This will allow the polymer to gel more completely. Mix again prior to use.

*Clearance*

Generally, gels should first be removed with the primary solvent of the gelled system. For paintings, it was found that the combined effect of solvents and the mechanical action of clearance achieves the best results [37]. Also note that although carbomer solvent gels do contain a small amount of water, attempts to clear them with aqueous systems will result in the precipitation of a hazy residue [38].

Again, even when carefully cleared, Ethomeen-Carbopol gels can leave residues on surfaces that are porous or uneven. With non-porous substrates, however, the gels were found to clear well, with the solvent proving more likely to leave residues than the polymer [18, 37].

***Preparation: Pemulen Gels***

Adapted from Stavroudis [39, 40] and Ravenel [41]

*Techniques and Tips*

**Caution**: No research about residues. Additional reagents and solvents may require PPE precautions. Do not allow gels to dry directly on the surface of the object.

**Lead time**: 2 days

**Useful concentrations**: 0.5–3 (wt. %)

**Formulation**: As previously discussed, Pemulen is technically a thickening emulsifier and not a physical gel. The preparation of Pemulen therefore more closely resembles the emulsification processes outlined in (Micro) Emulsions Section in Chapter 4. This is admittedly a stretch for this book's preparation-based groupings, but as Pemulen is primarily used as a rheology modifier rather than a surfactant, here it is. Those interested in developing more complex Pemulen systems are encouraged to see the section on (micro)emulsions in Chapter 4 and external resources such as the Modular Cleaning Program (MCP).

A quick note about process: There is a disconnect between the preparation methods described by conservators and those described by manufacturers. While this is no doubt partially due to questions of use and scale, there is potential for further research into new formulations.

In the stock solution method developed by Stavroudis for the MCP, and adapted in the following recipe, the Pemulen is dispersed directly into water, neutralised, and allowed to fully hydrate before the addition of non-polar solvents[40]. However, manufacturer brochures state that the opposite process is preferable, with Pemulen dispersed in room temperature oil before the addition of the neutralising water/alkaline solution [25].

**pH**: As with Carbopol, all pH readings for Pemulen solutions should be done with pH indicator papers, rather than a pH meter as the latter will give incorrect values. Although the exact mechanism has not as of yet been determined, it is thought that the polymer may interact with the reference electrode, distorting any results despite calibration [33].

See 'Technical Information' section for list of selected bases and ratios required for neutralisation.

*Materials*

Ratios
Simple Pemulen gel (pH 7–7.5):
100 mL deionized water
1.5 g triethanolamine (TEA)
1 g Pemulen TR-2

Pemulen non-polar solvent gel (pH 6–6.5): Please note that this ratio yields a 2% Pemulen gel which is then diluted to a final 1% w/v Pemulen solvent gel

50 mL deionized water
non-polar solvent
1 g triethanolamine (TEA)
1 g Pemulen TR-2
Weighing boats
Balance
Volumetric measure: graduated cylinder, syringe, etc.
Magnetic stirrer is ideal, but a balloon whisk will also work.
Container with well-sealing lid
Beaker: 50 mL or greater
pH paper (not a pH meter)
For solvent gels: Volumetric scoop (such as a tablespoon or 'cup' from cooking)

*Instructions*

1. Label the lidded container.
2. In the lidded container, prepare the Pemulen dispersion.
   (a) Measure out the deionised water and add roughly half to the container. Put the other half in a separate beaker and set aside until step 3.
   (b) If including any chelation agents or surfactants in your formulation, add these now.
   (c) Weigh out the Pemulen.
   (d) Stir the water vigorously and slowly dust the Pemulen powder on the surface.
   (e) Mix until the Pemulen is well dispersed and then set aside.
3. In the separate beaker, prepare the alkaline solution.
   (a) This beaker should contain the other portion of deionized water.
   (b) Weigh out the alkali, add to the water and mix until dissolved.
4. Using a magnetic stirrer or balloon whisk, begin mixing the Pemulen solution.
5. Slowly add the full quantity of alkaline solution to the Pemulen dispersion. Stir until very well mixed.
6. Set the gel aside for several days. This allows the Pemulen to more fully react with the alkali and the pH to stabilise.
7. After the rest period, check pH by diluting the Pemulen mixture with deionized water until thick liquid is formed (approximately 4 mL water per gram of Pemulen gel). Check pH with litmus papers rather than a meter, and adjust by adding further Pemulen or alkaline solution as needed. If creating a simple gel, this is your final step.

8. If creating a solvent emulsion:
   (a) Decant a small amount of the gel, noting the volume, and place in the second lidded container. Standard scoops (or tablespoons) can be helpful here.
   (b) Measure out an equal *volume* of non-polar solvent and add to the container.
   (c) Mix well using a magnetic stirrer or small balloon whisk until a transparent, creamy-feeling emulsion has formed.

*Clearance*

There has been no specific research into the quantity or characteristics of residues which may be left behind by Pemulen systems, but preliminary discussions are broadly encouraging [42]. As ever, however, absence of evidence is not evidence of absence, and further research is required.

***Preparation: Methyl Cellulose***

For full discussion of preparation methods, see the section 'Cellulose Ethers' in Chapter 2.

*Techniques and Tips*

**Caution**: MC is non-toxic, but added reagents may require PPE precautions. Potential for residues. Do not allow gels to dry onto the surface of the object.
**Lead time**: several hours
**Useful concentrations**: 5–10% (w/v) methyl cellulose to water
**Formulation**: Methyl cellulose can be supplied in more than a few different molecular weights and viscosities. See the technical data section for a selection of these figures and Cellulose Ethers Section in Chapter 2 for additional details and preparation methods.
**Mixing**: Generally, stirring is preferred over shaking, as the latter can introduce air bubbles which can affect gel formation and the functionality of some reagents (particularly enzymes). However, with simple aqueous solutions and in the early stages of preparation, shaking the container can help disperse the MC particles. If attempted, ensure that the containers are very well sealing and remember that this should not be attempted with volatile solvents or caustic reagents. Stirring, particularly with a magnetic stir bar or handheld milk frother, is by far the preferred method.

*Materials*

Cellulose ether powder
Primary solvent (typically water or alcohol)
Container with tight-fitting lid

Graduated cylinder
Beaker
Watch glass or weighing paper
Whisk or milk frother
Optional: solvent
Optional: chelation agents, enzymes, buffers, pH adjustment. If using enzymes, ensure that lab equipment is compatible. See 'Enzymes' in Chapter 4.

*Instructions*

1. Label the lidded container with the materials and concentrations.
2. Use the graduated cylinder to measure out the total desired amount of water. If adding additional cosolvent, be sure to take this volume into account.
3. Pour roughly half of the water into the lidded container
4. If using other water-soluble additives (chelation agents, buffers, etc.), weigh or measure out each type separately and add to the water in the lidded container. Do not add enzymes at this point; see step 15.
5. Using the watch glass or weighing paper, weigh out the desired amount of CE solids and set aside. CEs are typically very light and can be easily scattered by air flow or swift movement. Cover with an extra watch glass if required.
6. Mix well.
7. Add the powdered CE to the container and mix for at least one minute.
8. Stop mixing and allow solution to rest for a few minutes.
9. Add the rest of the water and, if necessary, check and adjust pH.
10. Add any remaining co-solvent and/or co-solvent-surfactant mixtures (see '(Micro)emulsions').
11. Mix again until smooth and homogenous.
12. Visually inspect mixture for any clumps of CE power; break these up with a gloved finger or implement.
13. Set aside to continue to thicken. This can take anywhere from 20 minutes to several hours.
14. If mixture remains too runny after resting, additional MC can be added and the mixing repeated.
15. If using enzymes, use a water bath, sous vide, or other indirect warming method to warm the gel to the activity temperature range for the enzyme selected. Add the enzymes to the mixture and carefully stir. Avoid the inclusion of air as this can cause enzymes to denature.

*Clearance*

Methyl cellulose gels have not been as historically problematic as other thixotropic systems, but careful clearance is still required. In paintings conservation, two rinses with compatible solvents have been found to be sufficient, but again uneven or complex topography will require further rinsing or use of a barrier layer [35]. The first solvent should be the same as, or compatible with, the gelled system, while the second rinse should be compatible with the first, but able to clear away any excess materials. Water and white spirit are common choices.

***Preparation: Xanthan Gum***

*Techniques and Tips*

**Caution**: XG is non-toxic, but added reagents may require PPE precautions. Do not allow gels to dry onto the surface of the object.

**Lead time**: a few hours

**Useful concentrations**: 0.5–2% w/v

**Formulation**: Xanthan gum is a non-toxic and easy-to-use rheology modifier and emulsifier that has come into increased use in household kitchens and conservation laboratories alike. This growing popularity has had the effect of driving up accessibility, driving down prices, and introducing a huge range in quality. To what extent this variability may impact conservation gels has yet to be determined.

In XG systems, aqueous cleaning solutions can fully prepared, with all surfactants added and pH adjustments complete, before being easily gelled with the addition of xanthan gum [26]. As discussed previously, a 2% XG gel can support the addition of up to 20% (w/w) of non-miscible solvent or 40–50% (v/v) of miscible solvent [21].

Xanthan gum mixtures will continue to thicken upon resting. It is therefore recommended to allow the mixture to rest several hours before use.

**pH**: Xanthan gum is pH neutral and can form gels and emulsions at most pH values likely to be encountered in conservation treatments (2–12) [21, 27].

*Materials*

Ratios:

2 g xanthan gum

80 mL deionised water

20 mL solvent for the solvent gel or 20 mL deionized water for aqueous cleaning gels.

Please note: These ratios yield 80 mL of 2.5% (w/v) XG gel, which is then diluted to a final 100 mL 2% (w/v) solvent or aqueous gel.

Volumetric measure

Container with lid

Weighing paper or watch glass

Balance

Stirrer: magnetic, milk frother, or whisk (not for enzyme gels), or glass stirring rod.

If using enzymes, ensure that the lab equipment is compatible. See the section on enzymes in Chapter 4.

*Instructions*

1. Label the lidded container with the materials and concentrations.
2. Use the graduated cylinder to measure out the water and add to the container.
3. Using the watch glass or weighing paper, weigh out the desired amount of XG solids. As with other fine powders, XG powders can be very light, and can be easily scattered by swift movement. Cover with an extra watch glass if required.
4. Returning to the lidded container, stir the water gently, creating a small vortex.
5. While stirring, slowly sprinkle the XG powder on the surface.
6. Continue to mix for at least one minute. Visually inspect mixture for any agglomerates of XG power; some small ones are okay, but larger clumps should be broken up with an implement or gloved finger.
7. Stop stirring and set aside. The mixture will continue to thicken over the next hour or two. This is now a 2.5% stock solution, which can then be further diluted by the following methods. For simple aqueous cleaning systems, see Step 8. If creating a solvent gel, proceed to Step 9, and if using enzymes skip directly to Step 10.
8. For aqueous cleaning gels

   (a) Use a graduated cylinder to measure out additional water.
   (b) Measure out and add any desired solids (pH modifiers, surfactants) directly to the water and stir well.
   (c) Add this water to the XG gel and mix well. If the gel is a bit thick, a glass stirring rod or spatula might be easier than a magnetic stirrer.
   (d) Adjust pH as necessary with pH meter or indicator papers.

9. For solvent gels

   (a) Use a graduated cylinder to measure out additional solvent(s).
   (b) Add any miscible solvents and stir well.
   (c) Allow gel to rest a few minutes while you measure out any non-miscible solvents
   (d) Add any non-miscible solvents and stir well.
   (e) Allow gel to rest for several hours prior to use.

10. For enzyme gels:

    (a) Adjust the pH of the gel to within the desired range for the specific enzyme used.
    (b) Measure out the deionized water and pour into mixing container.
    (c) Measure out and add the enzymes to the water, stirring gently.
    (d) Add the enzyme solution to the gel and stir carefully until well mixed. Avoid the inclusion of air as this can cause enzymes to denature.
    (e) Check and adjust the pH. Avoid overmixing where possible.
    (f) Use a water bath, sous vide, or other indirect warming method, warm the XG gel to the activity temperature range for the enzyme selected.

**Technical Data**

*Table 3.2* Polymerisation and characteristics of some rheology modifiers

| *Gel (Brand name)* | *Polymerization solvent* | *Formulation pH range* | *Type* | *Viscosity, (cP)* | *Average MW (Daltons)* | *Source* |
|---|---|---|---|---|---|---|
| Carbopol 934 NF | Benzene | 5.5–8.5 | Polyacrylic acid | 30,500–39,400 (0.5 wt% at pH 7.5) | 3,000,000 | [23, 43] |
| Carbopol 974P NF (Suggested replacement for 934) | Ethyl acetate | 5.5–8.5 | Polyacrylic acid | 29,400–39,400 (0.5 wt% at pH 7.3–7.8) | | [44, 45] |
| Carbopol 940 NF | Benzene | 5.5–8.5 | Polyacrylic acid | 40,000–60,000 (0.5 wt% at pH 7.5) | 4,000,000 | [23] |
| Carbopol 980NF (Suggested replacement for 940) | Ethyl acetate -cyclohexane cosolvent | 5.5–8.5 | Polyacrylic acid | 40,000–60,000 (0.5 wt% at pH 7.5) | 4.5 billion (maximum estimate) | [23, 30, 45] |
| Carbopol 941 NF | Benzene | 5.5–8.5 | Polyacrylic acid | 4,000–10,000 (0.5 wt% at pH 7.5) | 1,250,000 | [23] |
| Carbopol 981NF (Suggested replacement for 941) | Ethyl acetate-cyclohexane cosolvent | 5.5–8.5 | Polyacrylic acid | 4,000–10,000 (0.5 wt% at pH 7.3 – 7.8) | - | [45, 46] |
| Carbopol Ultrez 21 | Ethyl acetate -cyclohexane cosolvent | 5–11 | Polyacrylic acid | 50,000–60,000 (0.5 wt% at pH 7.5) | - | [47, 48] |
| Pemulen TR-1 NF | ethyl acetate-cyclohexane cosolvent | 4–8 | acrylic acid and C10-C30 alkyl acrylate crosslinked with allyl pentaerythritol | 14,000–26,500 (1.0 wt% at pH 7.5) | - | [24, 25] |

(*Continued*)

*Table 3.2* (Continued)

| *Gel (Brand name)* | *Polymerization solvent* | *Formulation pH range* | *Type* | *Viscosity, (cP)* | *Average MW (Daltons)* | *Source* |
|---|---|---|---|---|---|---|
| Pemulen TR-2 NF | ethyl acetate-cyclohexane cosolvent | 4–8 | Acrylic acid and C10-C30 alkyl acrylate crosslinked with allyl pentaerythritol | 4,500–13,500 (1.0 wt% at pH 7.5) | - | [24, 25] |
| Methocel A4M | water | 2–13 | Methyl cellulose | 4000 2% in water | 41,000 | [49, 50] |
| Methocel A4C | water | 2–13 | Methyl cellulose | 400 2% in water | 86,000 | [49, 50] |
| Xanthan gum | water | 2–12 | polysaccharide | 1550 1% in water | 2–7.5 million | [27, 28, 51] |

Molecular weights of Carbopol and Pemulen are tricky to define due to their complex branching and cross-linking character. The results given are from scientific databases and suppliers websites, but these are likely very approximate at best.

*Table 3.3* A compilation of some uses of spreadable gels

| *Gel* | *Primary Solvent* | *Secondary solvents* | *Additives* | *Source* |
|---|---|---|---|---|
| Carbopol | 70% (v/v) Acetone, ethanol, isopropanol, mineral spirits, xylene, MEK, pyrrolidinone, petroleum ether | Xylene Benzyl alcohol | pH adjustment: TEA, EDTA, DTPA, hydrogen peroxide, sodium perborate non-ionic surfactant: Brij 35, Canasol | [35, 42, 52–55] |
| Pemulen | Water | 50% (v/v) Benzyl alcohol, ethanol, isopropanol, pyrrolidinone, petroleum ether, xylene | Deoxycholic acid, Citric acid, Bicine, EDTA | [26, 42, 55] |
| Methyl cellulose | water | <2% 1-methyl-2-pyrrolidinone | pH adjustment: deoxycholic acid, citric acid, HEDTA, TEA, non-ionic surfactant: Triton X-100, 85, Brij 35 enzyme: lipase, pepsin, amylase other: | [4, 35] |
| Xanthan Gum | water | 20% (v/v) of non-miscible solvent: ethanol, isopropanol, benzyl alcohol, acetone 40–50% (v/v) of miscible solvent: xylene, white spirit, | pH adjustment: concentrated sulfuric acid, triammonium citrate, citric acid, potassium nitrate, Trisma DTPA, di-EDTA, tetra-EDTA, desferal, TEA, non-ionic surfactant: enzyme: lipase Other: thiourea | [53, 56–59] |

*Table 3.4* Neutralisation ratio for Pemulen and Carbopol gels (not comprehensive)

| *Base* | *Ratio (Base:Carbopol) (wt)* | *Ratio (Base:Pemulen) (wt)* |
|---|---|---|
| Sodium Hydroxide NaOH (18%) | 2.3: 1 | 0.5 |
| Ammonium Hydroxide Ammonia (28%) | 0.7: 1 | 0.3 |
| Potassium Hydroxide KOH (18%) | 2.7: 1 | 0.5 |
| Triethanolamine TEA (99%) | 1.5: 1 | 2.0 |
| Ethomeen® C-25 | 6.2: 1 | - |

Further examples available through manufacturer documentation [25, 60]

## NON-SPREADABLE GELS

Consultant: Sophie Barbisan

### *Material Information*

Precision Required: Low

*Overview*

For the sake of precision, there is a temptation to wade into the taxonomic mire that is 'gels'. Unfortunately, there doesn't seem to be a single term that fully encompasses the characteristics or behaviours of the many gel-like materials used in conservation. All more accurate descriptions of these materials, be it hydrogel, rigid, flexible, high-density polymetric dispersions, necessarily exclude other compounds with similar uses and handling properties. In the interest of understandability over precision, I will therefore be lumping together many chemically distinct but practically similar materials (and mixtures thereof!) under the heading of non-spreadable gels. Now, where there's a will there's a way, and these materials can be crumbled or stretched, but ideally this title highlights their cohesion, their preparation and use, and sets them in opposition to the truly smearable thixotropic gels discussed previously.

The use of gels in conservation is very much an evolving discipline, with a steady stream of research into new applications and materials. Given this breadth, the materials covered in this section include only two representative rigid polysaccharide hydrogels: agar/agarose and gellan; as well as two borax-modified flexible 'gels': konjac-borax and PVOh-borax. One notable omission is the entire suite of Nanorestore p(HEMA)/PVP hydrogels developed specifically for heritage conservation [61]. Although commercially available, well attested, and by all accounts very user friendly, they are supplied as premade gel sheets and therefore require no preparation guidance [62]. Those interested in the topic are strongly urged to check the publications page of the NANORESTART (NANOmaterials for the REStoration of works of ART) project website and/or invest in a copy of *Gels in Conservation of Art* (2017).

RESIDUES

As with everything in this chapter, the use of gelled systems introduces the possibility of residues on the object's surface. It's also worth mentioning a short aside about the risk of tideline formation within porous substrates. For although not technically residues, they are unsightly for all objects, and for paper surfaces such tidelines can encourage the preferential deterioration of the cellulose network [63, 64].

In terms of residues deposited by the gels themselves, the porosity, character, and topography of the substrate will influence both the amount of material deposited as well as how problematic such residues may be. Agar has been found to leave trace residues on paper, but no more than the surfactants also included therein, and PVA-B gels appear to easily peel away from Carrara marble [65, 66]. Several studies show that gellan gums can leave residues on paper surfaces and newly published research also points out changes in wettability for samples once treated with gellan [65, 67]. There is good indication that the use of a barrier layer, particularly Japanese tissue, can mitigate such deposition, but this is a subject of evolving research and careful evaluation of the surface is required prior to treatment.

Finally, as this is a topic of great potential and swift growth, I'd like to interject an additional word of caution. At the moment, there is simply not enough literature on the long-term effects of gel treatments to make definitive statements about their suitability. What work has been done urges further caution, particularly when working with complex polysaccharides, and I would extend this to include newly developed approaches. If the work by Henniges et al. and Sullivan et al. teaches us anything, it's that the issue of residues might not be a question of whether there are any, but instead of our ability to detect how much there is [65, 67].

### *Characteristics*

#### WATER FLOW

Gels are typically used to either hold a controlled volume of cleaning solution in close contact with a surface or to solubilize and then wick undesired materials up into the body of the gel. The water within hydrogels can be found in one of (approximately) two levels of availability. First there is 'bound' water that is chemically or physically confined to the gel network, unable to move around or interact with other molecules [68]. The 'free' water, as you might expect, flows easily through the pores and is available to react with or transport other molecules. The flow of water within the gel system, how and where it moves, and what it can carry is in part dictated by the size of the pores in the gel network.

With hydrogels like those made from gellan and agar, smaller pore size promotes capillary action that can preferentially pull material into the gel rather than releasing the water into the substrate [69]. Conveniently, the pore size of agar and gellan gels can be specifically controlled through the simple manipulation of component parts. In polysaccharide hydrogels, the higher the concentration of gelling agent to water, the more rigid the gel and the smaller the pores. Conversely, lower concentrations yield wetter-feeling gels with larger pore sizes. For gellan gels, this can also be linked to the form of gellan used (high vs low acyl) and the concentration of network-stabilising cations (usually calcium) [70, 71]. Broadly speaking, it's also worth noting that that water was found to release more quickly from agar/agarose gels than gellan [72, 73].

All that said, if taken in isolation this can paint an over-simplified picture where pore size alone dictates the movement of water within a gel. Unfortunately, such systems are rarely so simple. Although smaller pore size can encourage capillary action, it is only one factor among several which collectively drive the flow of water. Other factors include capillary pressure, osmotic pressure, gravity, and diffusion. As these elements are more challenging to account for than simple concentrations, there have, as of yet, been no specific guidelines developed for these factors. Instead, this is offered as a nod towards complexity rather than helpful tips for direct planning.

If you'll permit one more aside on pore size, however – although higher concentrations can release less water into the substrate, it doesn't follow that smaller is always better. Cleaning systems utilizing larger molecules (such as enzymes) or those that rely upon conformation changes (like some chelation agents) need to provide enough space for these molecules to effectively move, shift, and change, and therefore require roomier pores. For example, Bertasa et al. (2021) found that triammonium citrate exhibits more efficient complexing than EDTA when mixed into an agar gel, and surmises that this is likely due to the relationship between chelate and pore size even if the exact mechanism not yet determined [74].

### SYNERESIS

Finally, a quick note on syneresis. Syneresis is the withdrawal or shrinkage of a gel system coupled with the sudden expulsion of a liquid [21, 75]. Perhaps the easiest example is the puddle of water/whey that forms on the surface of a pot of yoghurt [76]. And while this water can be reincorporated though stirring your yoghurt, in conservation treatments, this sudden uncontrolled release can be detrimental or even disastrous. Rigid hydrogels, particularly agar gels, are known to exhibit this behaviour, particularly when weighted down to apply closer conformation [21]. Gel samples should therefore be subjected to treatment conditions and observed for similar behaviour.

### ADDITIVES

Additional reagents, enzymes, and solvents can be added to gel systems in one of several ways. For a more homogenous gel, compatible materials can be added directly the gum-water mixture and incorporated into solution prior to gelation. There is also the option to 'load' solvents and additives into the gels after they have formed. To do so, cooled gels can be immersed in baths containing the desired reagents and simply left for a few hours (or overnight), allowing the introduced components to diffuse into gel network. The suitability of each approach will depend on the exact combinations of gum and additives used. See 'Technical Information' section for further information.

Reagents can also be brushed or poured onto the surface of a gel immediately prior to use. This has proven a particularly useful application method for enzyme loaded gels, particularly given how finicky, fragile, and expensive these materials can be [77].

Finally, there is the promising, but sparsely explored possibility of drying, or freeze-drying, a gel and then rehydrating it with the desired solution [78]. Russick, Dobrowolski, Nosek, and Williams have had success with dried and rehydrated gellan gums, and food science publications regularly reference the freeze-drying of gellan to create loadable aerogels [78, 79]. See the section 'Freeze-Drying' for examples of this process.

#### SOL-PHASE

The point at which the gel particles have been dispersed in water but have not yet networked is sometimes referred to as the 'sol-phase' [80]. This reflects the colloidal nature of these mixtures and can be encountered in the technical literature. However, for clarity (perhaps at the cost of true precision), the general term 'dispersal' is preferred in this text.

### *Composition*

#### AGAR/AGAROSE

First, to sort out a few terms. Agar-agar is a mixture of two polysaccharides, agarose and agaropectin, extracted from the seaweeds *Gelidium* and/or *Gracilaria*. These two agar-prefixed compounds are the source of the double name 'agar-agar'. Of these two, only agarose exhibits gel-forming behaviour, and is therefore often isolated from the agar-agar mixture to provide a less complex and more homogeneous gel [81]. While perhaps less precise, the term agar-agar seems to be slowly being replaced by the abbreviated 'agar' and both singular and double forms of the term are used interchangeably [21, 49]. To summarize, agar-agar, now often called by the singular agar, consists of a less-refined mixture of both agarose and agaropectin.

Agarose is primarily composed of β-d-galactose and 3,6-anhydro-α-l-galactose, although there can be some variation in specific substitutions based on the exact seaweed source and the season of harvest [81]. Agaropectin has the same galactose backbone as agarose, but contains higher sulphur content and some pyruvic acid residues. These features prevent the formation of double-helices, which are vital to the gelling process [81].

The intertwining of these helices, the mechanism of which is still subject of some discussion, is also responsible for the formation of pores [21, 81]. The size and regularity of these pores directly influences the movement of water within the gel system. Furthermore, pore size is tightly correlated to gum concentration, with higher ratios of gum to water yielding gels with smaller pore size. This trend is attested for both agar and agarose gels, but the consistency of these pores varies with the agar source. The more refined agarose gels have been shown to exhibit pores of a near uniform size, while that of food-grade agar was found to be highly inconsistent [82]. As ever, how much of an issue this is will depend on the sensitivity of the substrate. If tidelines are a major concern, a more tightly controlled water flow might be preferable.

The molecular weight of agarose is 125,000 and it will set to a gel at concentrations above 0.7% [49]. Unusually for such gels, the melting point of agar and agarose is much higher than the gel point. The solution will initially set when cooled to 32–37°C, but requires reheating to the much higher 88–94°C to fully remelt [21, 49].

### GELLAN

Gellan is another polysaccharide, but rather than being extracted directly from a plant source (such as seaweed), gellan gums are produced as a by-product of the fermentation of sugar performed by *Sphingomonas elodea* bacteria [21]. This bacteria-mediated fermentation produces one type of 'high-acyl' (HA) gellan, which can then be further refined into the 'low-acyl' (LA) type. Both types exhibit unique gelling properties and characteristics, and both are commercially available. Just a quick aside to provide a functional group review: acyl groups are those which contain a carbonyl (C = O) bonded to carbon. In Gellan gum, high-acyl gellan has periodically alternating, acyl-containing, pendant glycerate, and acetate groups. Unsurprisingly, in the low-acyl form these side groups have been stripped away [21]. Practically speaking, high-acyl gellan yields flexible, opaque, semi-elastic gels, while low-acyl are rigid, clear, and brittle [83].

The gel networks of gellan gums are stabilised by the presence of mono and divalent cations (particularly the latter), which is often supplied in the form of calcium. While such additions aren't strictly necessary, rigid low-acyl gels can be formed at higher concentrations (2–5%), but these gels have comparatively poor stability and can be crumbly and difficult to handle [65]. This is especially true for low-acyl gellan, but high-acyl gels have also been found to be stabilised by the addition of calcium or magnesium salts [84].

Within the temperature range of most conservation activity, gellan gels are mostly thermo-**ir**reversible, and they can be warmed, cooled, or frozen without significantly affecting the gel character [21, 79]. Like agar gels, gellan gel networks are ionically stabilised and can therefore tolerate only a reasonably narrow, mildly acidic pH range of 3.6–6.5 [85]. They can, however, withstand the inclusion of enzymes, some chelating agents, surfactants, and some water-miscible solvents.

### KONJAC

Konjac is an old material, having been used in east Asian cooking for millennia, but it is a relatively new addition to the conservation tool kit. Konjac, also called by the Japanese transliteration 'konnyaku', is a white powder ground from the root of the *Amorphophallus* konjac plant. There are roughly 170 separate species of konjac plants, but this variety of sources does not appear to affect the gel properties of the material in a significant manner [86]. It is not, however, the entire konjac itself, which is used in konjac gum, but instead the glucomannan polysaccharide that is isolated from the starches, lipids, proteins, and other materials found within the plant. For this reason, konjac is also referred to a konjac glucomannan gum [87, 88].

Another polysaccharide, the main chain of konjac glucomannan consists of d-mannose and d-glucose with a few side acetyl groups. Like the acyl pendants of gellan gums, these acetyl groups are a key part of the gelling mechanism. The addition of an alkali material (usually borax) will deacetylate these chains and allow for the formation of non-thermo-reversible gels [88].

### POLYVINYL ALCOHOL

Polyvinyl alcohol is a popular polymer known widely as PVA, PVAL, and PVOH. A final quick note for those in need of functional group revision, alcohols, as a class of organic materials, are characterised by the presence of a pendant group of oxygen and hydrogen. PVOH is therefore indicating the presence of the poly, the vinyl, and the alcohol (–OH).

Although all three abbreviations are common, the latter two attempt to differentiate polyvinyl alcohol (PVOH) from polyvinyl acetate (PVAc). To compound the confusions, please note that both PVOH and PVAc can be modified with borax to produce flexible gels, and both are discussed in the conservation literature [89–91].

The distinction between PVOH and PVAc is an important one, but the reality is that any polyvinyl alcohol will contain some polyvinyl acetate. This is because PVOH is derived from PVAc and trace acetate groups, sometimes only very few, manage to survive the processing. As a result, different PVOH grades are referred to by their 'hydrolysis percent', or the relative amount of acetate groups still present. This ranges from partially hydrolysed (10–15 mol% acetate), to intermediate (3–7 mol%) and fully hydrolysed (1–2 mol%) [92]. This means that 99% hydrolysed PVOH, should contain only about 1% acetate (mol%), while 45% hydrolysed will contain roughly 45% PVOH to 55% PVAc [93].

PVOH polymer is based on the repeating 'vinyl group' ($-CH = CH_2$) with a hydroxide side chain. The vinyl double bond is lost upon polymerisation, but results in the formation of a high-molecular-weight hydrocarbon backbone. Some characteristics of PVOH are outlined in Resins into Solvents Section in Chapter 2 on adhesives, but it can also be modified by the addition of sodium tetraborate decahydrate (borax) to form flexible gels. When combined with borax, the borate ions and boric acid pluck the hydrogen from the alcohol hydroxide group, leading to the formation of a network of ester crosslinks between polymer chains [94]. Although their use in conservation predates their wider popularity as a home-craft, PVOH-borax gels will no doubt already be familiar to many as the toy 'slime'.

#### BORAX

Borax is not a polysaccharide, or a gelling agent, but as it is instrumental to the formation of several gels, it's worth a short mention here. Borax is the common name for water-soluble sodium tetraborate decahydrate salt with the formula $B_4O_7Na_2$ $10H_2O$ and a pH near 9.3. Borax is naturally occurring as an evaporate, left behind in dry lake beds and has been broadly used as everything from a pesticide to a flux for welding [95].

Borax is also a strong irritant, harmful to aquatic life, and a suspected mutagen [96]. All relevant PPE and precautions should be taken if working with borax.

### ***Preparation: Polysaccharide Gels***

*Techniques and Tips*

**Caution**: Possibility of residues
**Lead time**: 30 minutes to several hours depending on the application method and additives
**Useful concentrations**: Agar 1–10%; Gellan 2–5%

#### MATERIAL CHOICE

As discussed previously, even at stable concentrations, the pore character of agar, agarose, and gellan gums is all known to vary based on external factors such as refinement level (food grade vs reagent grade) or the presence of network stabilising cations.

A few quick notes about grade. In projects where the release of water does not need to be quite so tightly controlled, food grade agar might be suitable, but the ambiguity of composition leaves such domestic-grade materials open to further questions about residues. Similarly, for treatments where tap water can be tolerated, no additional calcium is required for the formation of gellan gums. Many municipal water services, particularly those in hard water areas, supply more than enough magnesium and calcium to stabilize both high- and low-acyl forms gellan, but tap water can also carry numerous, often undesirable, salts, metals, and other compounds [97].

#### RIGIDITY AND CONCENTRATION

Another factor in determining useful concentrations is the relationship between concentration and flexibility. Higher concentrations of agar or gellan correlates with gels which are 'dryer', but are also significantly more rigid. Rigid gels (around 5–10%) are less likely to mould to the contours of an object's surface, and even flat objects may require that the gel be weighted to maintain consistent contact. More flexible gels (1–5%) are usually much wetter feeling, quickly releasing water with simple handling. These will more closely conform to uneven surface topography but unless applied hot (see 'Casting'), will not form close contact with complex decorative surfaces.

One final reminder that the smaller pores found in higher concentration mixtures can constrain the activity of other molecules within the gel system. Larger molecules such as some surfactants and enzymes are most effective in gels with larger pores. Specifically, it was found that the activity of enzyme solutions was best supported by agarose gels with concentrations of 1–1.5% [77]. It's thought that the larger pore size and greater amount of free water in lower-concentration gels allows the enzyme molecule to move freely within the network [77].

### MIXING

When dispersing the gum in cool water, the solid particles will tend to clump. A magnetic stirrer will work moderately well, but I've found a small whisk or handheld milk frother to be the easiest and best way to carefully disperse the powder. A glass stirring rod also works well for mixing the liquid and breaking up any clumps.

If using gellan, the low-acyl form will disperse more quickly in the presence of calcium, while high-acyl doesn't disperse so much as it swells [21, 84]. That said, for whatever reason, gellan clumps more readily than agarose and is more difficult to disperse. It's therefore worth dusting it into solution at a much lower rate.

You can also dust the powder on water in a well-sealing container and shake it, pausing to break up any clumps with a gloved finger. However the simplicity of this method makes for a tempting technique, but it can result in uneven mixing and an inconsistent gel. Please note, this is of course only for the initial cold dispersal and should not be attempted with hot solutions.

The point at which the gel particles have been dispersed in water but have not yet networked is sometimes referred to as the 'sol-phase'. For clarity, the general term 'dispersal' is preferred in this text.

### HEATING

Dissolving the solid gum powder requires heat, and this can be done by either direct (hot plate, microwave) or indirect (sous vide, water bath) means. When making single batches for one-off projects, a hot plate or microwave works well, but if preparing several different mixtures at the same time, it's worth considering either the sous vide or bag-boil methods described below.

If you have a sous vide or hot plate large enough to support a water bath (such as those used in dye-making), you can pour the various gel mixtures into well sealing, heat-safe jars and place these in the bath. The water level should exceed the liquid in the jars but ideally fall short of the lid.

If you don't have a sous vide, you can essentially treat the mixture like boil-in-the-bag food. Label a series of nylon or polypropylene boil bags, prepare the gum dispersal as described in Steps 1–7 of the instructions, and then pour the mixtures in. Squeeze out the excess air and seal well, but also try to avoid overfilling the bags, leaving plenty of room for the liquid to slosh around inside. Place all the bags in a large beaker and fill halfway to two-thirds full of water. You want to cover the liquid in the bags (which may want to float) while also leaving enough space for the water to boil. At least once in the cooking process using heat-proof gloves, oven mitts, or a thick towel, pull each bag from the boil and mix the contents by lightly jostling the bag. Be *careful and gentle* when handling the bags as hot gels can easily burn skin.

Instructions for microwave heating are discussed in the following 'recipe', and while this method may seem a bit ad hoc at first look, it's actually recommended by both practitioners and manufacturers [69, 83, 98, 99]. Please note that at least agarose mixtures (if not all gums) can become super-heated in the microwave and can foam up suddenly when agitated. Be sure to handle hot mixtures with care and regularly stir mixtures following periods of heating [99]

Finally, if a more transparent agar gel is desired, its reported that this can be attained by heating the mixture twice [100]. As agar is fully thermo-reversible, it can be heated, allowed to cool and gel and then heated again.

#### CASTING

Once heated the gels will need to be either cast onto a flat surface or applied hot directly to the object surface. As for this latter technique, there is good precedent for this approach, and hot gel mixtures can be poured, extruded from a syringe, brushed, or even sprayed, directly on the object's surface [57, 101, 102]. However, it goes without saying (but I will anyways) that this obviously requires good knowledge of both the substrate and the soiling and is unsuitable for many surface types. For recent (and open access) examples of these techniques, see the projects undertaken by Giordano, Caruso, and Lazzara (2022) and Guilminot (2023) [57, 102].

A much more common approach is for the gel to simply be cast onto a smooth surface, allowed to cool and then cut into smaller pieces for further modification or direct use. Pretty much anything smooth, inert, and clean is acceptable for this: from silicon ice trays to non-stick baking sheets or the lids of Stewart polyethylene boxes (a personal favourite). If you're looking to weight the gels on the object surface, you can encourage uniform thickness by pouring out the hot mix, covering it with Melinex, and placing a board on top.

Large, thin sheets of rigid hydrogels can be difficult to handle, quickly breaking or tearing, due to their slippery and brittle character. Russick et al. have found that embedding a sheet of spun polyester (Reemay, Bondina, etc.) into one side of gellan gum supports handling, particularly with large castings [78].

Finally, remember that your casting surface is typically the side that will touch the object, particularly if using a polyester support. Consider smoothness, shape, and thickness when selecting casting surfaces. Furthermore, if intending to weight the gel to encourage even contact, ensure that the liquid is cast to a uniform thickness.

#### COMBINATIONS

The idea of combining different polysaccharide gellants has been informally discussed in tutorials, personal correspondence, and online forums like the Global Conservation Forum (aka the Dist List). At the point of writing, details have not been formally published, but Cushman has developed mixtures of xanthan, konjac, and agarose (x:k:a) with promising results [103]. It's reported that adjusting the relative concentrations can yield gels with very different properties. Of particular interest is a clear, flexible gel made from a ratio of 2:2:1 of xanthan:konjac:agarose with a total concentration 1–1.5% (w/v) in water [103, 104]. Again, however, formal research is not yet available, and all cautions about residues apply two-fold with novel materials.

### *Method: Microwave*

#### MATERIALS

Gum powder
Water: distilled, deionized, or tap, depending on project and any desired additives
Heat-proof glass beaker
Graduated cylinder
Watch glass or weighing paper
Balance
Microwave: ideally with working interior light
Glass stirring rod
Flexible spatula or wooden tongue depressor
Heat-proof glove or towel: strictly necessary. The gel will get very hot during heating and can quickly scald
Casting surface: non-stick baking tray, plastic container or lid, anything smooth and clean really.

INSTRUCTIONS

1. Prepare your casting surface next to the microwave.
2. Weigh out the desired amount of gum powder and set aside.
3. Using the graduated cylinder, measure out the desired amount of water and add to the beaker
4. Dust the gum powder on the surface of the water and mix until well incorporated, with no large clumps.
5. If incorporating any buffers, chelation agents, or other water-soluble reagents, add them now and mix well. Chelation agents can typically withstand the temperatures used here, but surfactants can affect surface tension, and therefore the boiling point, or can cause foaming. This isn't necessarily a problem, but it is good to be aware of.
6. Once the water and gum are well mixed, cover the beaker with a piece of cling film and pierce a small hole for ventilation.
7. Set the power level of the microwave to a mid-level setting to start. Begin heating the mixture bringing the solution to a low boil. Note: The beaker will get very hot during the cooking process, you will need to use something to allow you to safely handle it such as an oven mitt or several layers of folded paper towel
8. Boil for 30–60 seconds. Remove from microwave and stir well. Try to break up any clumps.
9. Return to the microwave, heating for another 30–60 seconds and then stopping and stirring again.
10. Repeat this another 3–4 times, resulting in a total cook time of between 2 and 5 minutes (depending on batch size and microwave power). In terms of temperature, simmering and boiling is fine, but prolonged vigorous boiling should be avoided as the mixture can scorch. You're looking for the gum powder to fully dissolve, leaving a uniform mixture with no clumps or obvious solids.
11. If using any co-solvents, add these now and mix well. Mind the fumes from the hot liquid and work with extraction if necessary.
12. Once ready, use the heat-proof gloves and pour the mixture directly onto your casting surface. Use a spatula or tongue depressor to scrape the excess out of your beaker. Depending on the concentrations used, the liquid can be fairly viscous or runny, but all should easily settle into a level plane.
13. Allow to cool. The time this takes will depend on the thickness of the gel and the temperature of your workspace, but 10–30 minutes is more than sufficient for most.
14. Cut the gel into the desired shape and size.
15. Rinse these pieces in water and pat dry with paper towel.
    Optional

16. If 'loading' further reagents, solvents, or surfactants into the gels, mix such solutions separately and then immerse the gel cubes within them. Allow to sit for an hour or two so the additional materials can migrate into the gel body.
17. If adding enzymes, warm the gel using an indirect warming method (sous vide, or place a dish in a pan of hot water) and heat until the gel reaches the desired activity temperature. Mix the enzyme solution separately and then brush or pipette it onto the surface of the warm gel. Repeat this several times and then apply the gel to the object with the enzyme-loaded face against the surface.

### *Method: Hot Plate*

#### MATERIALS

Gum powder
Water: distilled, deionized, or tap, depending on context and any desired additives
Heat-proof glass beaker
Graduated cylinder
Watch glass or weighing paper
Balance
Hot plate
Plastic wrap/cling film
Stirrer: magnetic or glass rod
Flexible spatula or wooden tongue depressor
Heat-proof gloves
Casting surface: non-stick baking tray, plastic container or lid, anything smooth and clean really.
Optional: solvent
Optional: chelation agents, enzymes, buffers, pH adjustment. If using enzymes, ensure that lab equipment is compatible. See the section on enzymes in Chapter 4.

#### INSTRUCTIONS

1. Weigh out the desired amount of gum powder and set aside.
2. Using the graduated cylinder, measure out the desired amount of water.
3. If using the 'seal and shake' method, follow the instructions outlined in Cellulose Ethers Section in Chapter 2 and skip to Step 7. If using a magnetic stirrer follow Step 4.
4. Add the water to the beaker and place on the hot plate.
5. Place the magnetic stir bar in the water and set to rotate on a medium-low setting. The water should be clearly swirling, but a vigorous vortex isn't necessary.

6. Dust the gum powder on the surface of the water and mix until well incorporated, with no large clumps.
7. If incorporating any buffers, chelation agents, or other water-soluble reagents, add them now. These additives can typically withstand the temperatures used here, but be mindful of ventilation and volatility, avoiding overheating ammonia-containing compounds. Surfactants can also affect surface tension, and therefore the boiling point, or can cause foaming. This isn't necessarily a problem, but it is good to be aware of.
8. Once the water and gum are well mixed, cover the beaker with a piece of cling film and pierce a small hole for ventilation.
9. Begin heating the mixture (either side of medium depending on your device), bringing the solution to a simmer or low boil. This temperature should be maintained until the gum dissolves. You're looking for a uniform mixture with no clumps or obvious solids. The timing depends on volume and concentration but either side of five minutes is sufficient for most smaller batches. In terms of temperature, simmering and boiling is fine but prolonged vigorous boiling should be avoided as the mixture can scorch.
10. While this is heating, prepare your casting surface.
11. Once the gum is dissolved and the mixture is homogenous, turn off the heat. If using any co-solvents, these can either be added here, or 'loaded' into the gel once is has cooled (see Step 16). When adding co-solvents to the warmed mixture, ensure that the temperature of the gel dispersion has dropped below the solvent's boiling point and mix well. Mind the fumes from the hot liquid and work with extraction if necessary.
12. Then, use the heat-proof gloves and pour the mixture directly onto your casting surface. Use a spatula or tongue depressor to scrape the excess out of your beaker. Depending on the concentrations used, the liquid can be fairly viscous or runny, but all should easily settle into a level plane. At least they should if both your workstation and casting surface are themselves level.
13. Allow to cool. The time this takes will depend on the thickness of the gel and the temperature of your workspace, but 10–30 minutes is more than sufficient for most.
14. Cut the gel into the desired shape and size.
15. Rinse these pieces in water and pat dry with paper towel to remove any superficial residues.

    Optional
16. If 'loading' further reagents, solvents, or surfactants into the gels, mix such solutions separately, and then immerse the cool gel cubes within them. Allow to sit for an hour or two so the additional materials can migrate into the gel body.
17. If adding enzymes, warm the gel using an indirect warming method (sous vide or place a dish in a pan of hot water) and heat until the gel reaches the desired activity temperature. Mix the enzyme solution separately and then brush or pipette it onto the surface of the warm gel. Repeat this several times and then apply the gel with the enzyme-loaded face against the surface being treated.

***Preparation: Borax-Modified Gels***

Methods adapted from Angelova et al. (2017), Al-Emam et al. (2019), and Lee et al. (2022) [93, 105–107].

*Techniques and Tips*

**Caution**: Possibility of residues, particularly with the Konjac Glucomannan–Borax mixture

**Lead time**: 30 minutes to several hours depending on the application method and additives

**Useful concentrations (w/v)**:

PVOH 4-14% Borax 1–2% (This depends on hydrolysis of PVA used. See 'Technical Information')

MIX: 3% of PVOH, 1% of agar, and 0.6% of borax

Borax-modified PVOH gels are stretchy, flexible, and self-healing, and while these features can make them difficult to cast or cut into specific shapes, they can also allow them to conform more closely to complex surfaces. For PVOH-B gels, the characteristics of the gel will depend on the hydrolysis percent (see 'Overview') and molecular weight of PVOH used, but these can also be modified by variations of pH or temperature [93]. Broadly speaking, less hydrolysed PVOH samples with a greater amount of acetate groups (40–75% PVOH) will tolerate solvents better than the more 'pure' PVOH (80–99%) [93]. A good middle ground seems to be found with 75% PVOH, such as Kuraray Poval 5-74, which tolerates up to 50% (w/v) of many common solvents [93]. See 'Technical Information' for more information.

Those looking to split the difference between rigid and flexible are encouraged to look into the PVOH-B and agarose blends developed by Al-Emam, Motawea, Janssens, and Caen (2019), and those interested in the behaviour of different solvents are referred to Tobin and Sawicki (2023) [105, 108]. Lee et al. (2023) have described a similar flexible gel made from borax-modified Konjac [107]. Although this research reflects a promising approach, it is a new technique and the potential for residues has not yet been explored. Casual experiments with food-grade reagents yielded a flexible gel, but also left residues on gloves and paper towels. This is no doubt worsened by the low-quality konjac used, but further research is required.

*Method: Polyvinyl Alcohol – Borax (PVOH-B) Gel*

MATERIALS

Polymer powder and borax powder
Water: distilled or deionized
2 heat-proof glass beakers
Graduated cylinder
Watch glass or weighing paper

Balance
Hot plate
Stirrer: magnetic or glass rod
Flexible spatula or wooden tongue depressor
Heat-proof gloves
Optional: Organic solvents
Optional: chelation agents, enzymes, buffers, pH adjustment

INSTRUCTIONS

1. Using the graduated cylinder, measure out the total desired amount of water.
2. Add most of the water to the beaker, leaving only a few millilitres to dissolve the borax.
3. Add the magnetic stir bar to the beaker, place it on the hot plate/stirrer and start the rotation on a medium-low setting. The water should be clearly swirling, but a vigorous vortex isn't necessary.
4. Dust the gel powder on the surface of the water and mix until well incorporated, with no large clumps. For PVOH powders, this might require gentle heating up to around 50°C to encourage dissolution.
5. While the gel powder is hydrating, add the other remaining water to the second beaker.
6. Weigh out the desired amount of borax and add this to the water. Stir well until the borax is dissolved. To speed things along, you can also heat the borax mixture to around 50°C.
7. Once the solutions are fully dissolved, allow both to cool to room temperature.
8. Return the gel solution of the mixer and turn the rotation to high. Add the borax solution to the gel solution. The gel should form almost immediately.
9. This can then be turned out onto a casting surface.

*Method: Konjac Glucomannan – Borax (KG-B) Gel*

*Caution: New technique with potential for residues*, particularly with food-grade reagents. See the techniques and tips section. Further research is required prior to use on objects.

Konjac: 1–3%; Borax 0.5–1.5% (w/v)

MATERIALS

Konjac powder and borax powder
Water: distilled or deionized
2 heat-proof glass beakers

1 larger heat-proof container, for heating the gel once formed
Graduated cylinder
Watch glass or weighing paper
Balance
Hot plate
Stirrer: whisk or glass rod
Flexible spatula or wooden tongue depressor
Heat-proof gloves
Optional: chelation agents, enzymes, buffers, pH adjustment

INSTRUCTIONS

1. Using the graduated cylinder, measure out the total desired amount of water.
2. Add most of the water to the beaker, leaving only a few millilitres to dissolve the borax.
3. Dust the konjac gel powder on the surface of the water and mix until well incorporated, with no large clumps.
4. While the gel powder is hydrating, add the remaining water to the second beaker.
5. Weigh out the desired amount of borax and add this to the water. Stir well until the borax is dissolved. To speed things along, you can also heat the borax mixture to around 50°C.
6. Once the borax is fully dissolved, allow to cool to room temperature.
7. Return the konjac dispersion and add the borax solution to the konjac solution. Mix well, but not over-long as the mixture should start to gel almost immediately.
8. Allow to rest for at least 30 minutes.
9. Prepare a double-boiler style water bath in the larger heat-proof container and heat to around 90°C (±10). The water should be high enough to surround the base of the solution-containing beaker.
10. Place the beaker with the konjac-borax gel into the water bath and allow it to heat for around 20 minutes.
11. The gel can then be turned out onto a casting surface with the excess water able to drain away..
12. If using cleaning solutions, prepare these (see Chapter 4) separately and then immerse the KG-B gel for several hours (around 5).
13. Remove from cleaning solution, and pat dry the gel to remove any surface liquid prior to use.

***Storage and Drying***

Unfortunately, gels are susceptible mould, particularly if left unrefrigerated. Although the viable storage duration will depend on the precise gums, solvents, and additives used, two weeks has been cited as a general maximum [98] It is possible that this might be extended by thoroughly drying or freeze-drying the samples, but this has not yet been formally demonstrated. However, informal tests have indicated that air-dried samples remained susceptible to biological growth within a few days, whereas freeze-dried samples appeared unaffected over a three week period.

That said, although the effects on longevity are unknown, it has been shown that both agar and gellan gels can withstand drying or freeze-drying processes without provoking structural collapse [79, 109–112]. In these instances, the free water is driven off from the network, resulting in a dry sponge-like gel. This 'sponge' can then be rehydrated with any appropriate solution.

The ability to precisely control concentration and rehydrate a gel with a homogenous solution carries obvious benefits. There has, of yet, been no conservation research into freeze-dried hydrogels, but it is an established technique in the food science and pharmaceutical industries [79, 109–112]. Although far short of a carefully controlled experiment, informal tests by the author have provided proof of concept within heritage conservation and these are summarized in 'Freeze-Drying'.

### *Air Drying*

Russick et al. were able to successfully air dry thin layers of 3% (w/v) gellan gum within a day or two [78]. The gellan cast was sandwiched between layers of Bondina (non-woven polyester textile) and placed between window screens and simply allowed to dry overnight [78]. The authors note that this might not be successful in more humid environments, and suggest the use of light warming, increased air flow, or a desiccation chamber. The resulting dry-gel is still somewhat flexible, but the shape is noticeably deformed. In terms of warming Cassanelli et al. (2018) have had success with oven drying of LA gellan gum, and have also explored the effects of pH on the uptake of water during rehydration [79, 113].

Tests by the author in springtime London (15 °C and 60–70% RH) found that the air-dried samples shrivelled, becoming thin and uneven, particularly at low concentrations. Following rehydration via immersion, these samples did regain some weight (roughly 20–50%), but remained thin and deformed. At lower concentrations (1%) it appears that the loss of water through slow evaporation leads to total collapse of the gel network. This was less apparent at 3–5% concentrations and although some curling and discolouration was noted, it was nowhere near as pronounced as with the 1% gel.

*Freeze-Drying*

As previously discussed, research from bio-sciences and the pharmaceutical industry has demonstrated that both gellan and agar gels can be successfully freeze-dried and rehydrated [79, 109–112]. There is, however, one potential hiccup. This method has been shown to affect pore size, with the large crystals formed in the freezing stage leading to larger pores in the dried network [113]. The effect this may have on water retention and therefore treatment has not been explored.

Results from testing show that freeze-dried agar and gellan gels lost about 90–95% of their weight, but retained their overall shape, showing none of the deformations of the other drying methods. The freeze-dried gels had the consistency of packing peanuts, very lightweight but not overly fragile.

Upon rehydration, the gels regained about 70–95% of their weight. However, visual examination found that that the pores for both gel types appeared much larger than control samples, appearing more like small sponges. The network also was largely sponge-like, with the gel samples easily compressed (particularly at lower concentrations). The effect this may have on syneresis and water retention was not explored but changes to both are implicated.

In summary, freeze-drying effectively removed water from the hydrogel without causing collapse of the network structure. All gels were easily rehydrated, regaining most of the lost water within an hour. Their consistency was affected by the freezing process, and this may have implications for treatment.

***Technical Data***

*Table 3.5* List of common gellants and select characteristics

| *Agar* | *MW* | *Gel temp* | *Hydration temp (°C)* | *pH range tolerance* | *Melt temp* | *Source* |
|---|---|---|---|---|---|---|
| Agarose/ Agar | 80,000–140,000 | 32–40 | 85 | 4.5–10 | 85 | [114–116] |
| High-acyl (HA) gellan | 1,000,000–2,000,000 | 70–80 | 85 | 3.6–6.5 | 70–80 | [84, 85] |
| Low-acyl (LA) gellan | 200,000–300,000 | 30–50 | 70–95 | <3.6–6.5 | Not usually thermo-reversible | [84] |
| Konjac | 200,000–2,000,000 | 85–90 | - | >9 | Not usually thermo-reversible | [106, 107, 117] |

*Table 3.6* A compilation of some uses of non-spreadable gels

| *Gel* | *Additive type* | *Additives* | *%* | *Use notes* | *Source* |
|---|---|---|---|---|---|
| PVOH-B | solvents | Methanol, ethanol, 1-propanol, 2-propanol, acetone, methyl acetate, n-methylpyrrolidone | 30–75% (v/v)<br>Up to 50% (w/w) | Generally, lower hydrolysis percent and therefore lower PVOH% corresponds to better tolerance of solvents.<br>Alcohols tolerated 40–99% PVOH.<br>Ketones tolerated only ≤ 75% PVOH | [93, 118] |
| | pH modifiers | None listed | | **Incompatible** with EDTA, TAC, nitric, citric and formic acids. | [119] |
| | surfactants | Ecosurf EH-6 | 1 (w/v) | Does not tolerate acids (nitric citric formic) | [93, 120] |
| | Microemulsions | Aqueous micellar and o/w microemulsions | | Loaded into gels. No specifics given. | [118] |
| PVOH-B / agar mixture | solvents | acetone, 1-butanol, 1-pentanol, methyl ethyl ketone, ethanol, xylene | 20% (v/v) | Although their inclusion was supported by the gel network, non-polar solvents are more likely and more quickly expelled | [121] |
| Agar | solvents | Ethanol, acetone-ethanol mix, Stoddard's | - | Added into dispersed gel (8% agar). Acetone alone and xylene found to destabilise gel network. 8% agar | [122] |
| | pH modifiers | TAC, EDTA,TEA, DTPA, phosphate buffer, potassium nitrate, sodium hypophosphite<br>Nitric acid ($10^{-4}$ M) and Oxalic acid (5 wt%), boric acid, citric acid, hydrogen peroxide, sodium hydroxide (5M) phosphoric acid | 0.5–2% | Small amounts of acids and bases (such as used for pH adjustment) are tolerated.<br>Higher concentrations of acids (5–10%) must be 'loaded' into cold gels, but are temporarily tolerated at best. Largely incompatible with nitric, citric and formic acids<br>Attempted to incorporate sodium hydroxide at concentrations of 2% (w/v) failed. | [57, 119, 123–125] |

| | | | | | |
|---|---|---|---|---|---|
| | surfactants | Tween 20 and other non ionic surfactants | 1% (w/w) | 1–4% (w/w) agar<br>Triton XL-80N and Orvus Paste found to destabilise gel network. 8% agar | [100, 122] |
| | microemulsions | water, sodium dodecylsulphate, pentan-1-ol, and petroleum ether | | 2% agar gel tolerates up to 1.5 % (w/w) petroleum ether. | [126] |
| | Bioactive | Bacteria<br>*Ochrobactrum sp.* and *Pseudomonas stutzeri* | | 1% agar. For removal of animal glue from paper. *Ochrobactrum sp* was 'entrapped' within agar.<br>2% agar. For removal of nitrate efflorescence. Here agar used to provide stable RH conditions for the *Pseudomonas* bacteria that had been separately applied. | [127–129] |
| | bioactive | Enzymes: α-amylase, trypsin and protease | | In 1–1.5% agarose | [130–133] |
| gellan | pH modifiers | DTPA, TAC, EDTA, potassium nitrate calcium propionate, calcium bicarbonate, borane tert-butylamine | | Many buffers retain pH when gelled, but some variation was found (particularly for DTPA buffered to pH 8 which dropped to 6.8, but most others in the +- 0.1<br>Largely incompatible with nitric, citric and formic acids | [69, 83, 98, 119, 134] |
| | solvents | Ethanol, methanol, isopropanol | | LA – gels made up and then immersed in solvents to 'load' the gels<br>HA – can be made up with organic solvents. | [135] |
| | Bioactive | Enzymes: α-amylase and proteinase | | 1.5–2% (wt) | [131, 132, 136–138] |
| | Bioactive | Bioactive molecules extracted from *Cnidaria* | | 2.5% (?) gellan gel | [139] |

*Drying Tests*

**Control**: Placed in refrigerator, covered. Duration: 36 hrs. Temperature: 4°C
**Air dried**: Between screens on bench top. Dehydration duration: 36 hrs. Temperature 15–20°C
**Freeze-dried**: Placed on plastic tray, uncovered. Dehydration duration: 24 hrs. Temperature –30°C
**Rehydration**: 1 hr: immersion in tap water

*Table 3.7* Tabulation of weight percent change of refrigerated, air dried, and freeze-dried samples of gellan and agar gels

| | *Dehydration: Weight loss (wt%)* | | | *Rehydration: Weight gain (wt%)* | | |
|---|---|---|---|---|---|---|
| *Gel % (w:v)* | *Control* | *Air dried* | *Freeze dried* | *Control* | *Air dried* | *Freeze dried* |
| Agar 1% | 1.45% | 91.07% | 96.28% | -7.19% | 41.51% | 73.58% |
| Agar 3% | 0.72% | 60.50% | 96.24% | -1.65% | 23.05% | 92.04% |
| Agar 5% | 0.72% | 58.70% | 89.34% | 0.48% | 23.28% | 79.54% |
| Gellan 1% | 0.65% | 76.06% | 97.85% | -2.53% | 31.53% | 94.57% |
| Gellan 3% | 0.80% | 45.07% | 96.30% | 3.85% | 29.63% | 95.21% |
| Gellan 5% | 0.22% | 56.85% | 93.76% | 4.71% | 47.32% | 92.95% |

## References

[1] S. Gänsicke, P. Hatchfield, A. Hykin, M. Svoboda, and C. M.-A. Tsu, "The Ancient Egyptian Collection at the Museum of Fine Arts, Boston. Part 2, a review of former treatments at the MFA and their consequences," *Journal of the American Institute for Conservation*, vol. 42, no. 2, pp. 193–245, 2003, doi: 10.2307/3180070
[2] C. Isaksson, "Solvent gel versus solvent poultice: evaluating two techniques for the removal of pressure-sensitive tape stains from paper," in *Gels in the Conservation of Art*, L. V. Angelova, B. Ormsby, J. Townsend, and R. Wolbers, Eds., London: Archetype, 2017, pp. 126–128.
[3] J. M. Huggett, "Clay minerals," in *Encyclopedia of Geology*, R. C. Selley, L. R. M. Cocks, and I. R. Plimer, Eds., Oxford: Elsevier, 2005, pp. 358–365. doi: 10.1016/B0-12-369396-9/00273-2
[4] S. Rivers and N. Umney, *Conservation of Furniture*. Abingdon: Routledge, 2013.
[5] C. Cleere, "L'Innocenza Perduta (Lost Innocence): conserving a carrara marble statue," *Journal of Conservation and Museum Studies*, vol. 2, 1997, doi: 10.5334/JCMS.2971
[6] PubChem, "Kaolin." Accessed: Aug. 13, 2023. [Online]. Available: https://pubchem.ncbi.nlm.nih.gov/compound/56841936
[7] R. L. Frost and J. Kristof, "Raman and infrared spectroscopic studies of kaolinite surfaces modified by intercalation," in *Interface Science and Technology*, vol. 1, F. Wypych and K. G. Satyanarayana, Eds., in Clay Surfaces, vol. 1, Elsevier, 2004, pp. 184–215. doi: 10.1016/S1573-4285(04)80041-3
[8] S. Jatav and Y. M. Joshi, "Chemical stability of Laponite in aqueous media," *Applied Clay Science*, vol. 97–98, pp. 72–77, 2014, doi: 10.1016/j.clay.2014.06.004
[9] B. K. Heragh, H. Taherinezhad, G. R. Mahdavinia, S. Javanshir, P. Labib, and S. Ghasemsolb, "pH-responsive co-delivery of doxorubicin and saffron via cross-linked chitosan/laponite RD nanoparticles for enhanced-chemotherapy," *Materials Today Communications*, vol. 34, p. 104956, 2023, doi: 10.1016/j.mtcomm.2022.104956

[10] J. Dinsmore, "Publication Contribution - Conservator's Cookbook," [Personal Communication] Apr. 19, 2024.
[11] C. Woolfitt and G. Abrey, "Poultices and the cleaning of historic masonry and sculpture," Building Conservation. Accessed: Aug. 26, 2023. [Online]. Available: https://www.buildingconservation.com/articles/poultices/poultice.htm
[12] L.-M. Lee, P. Rogers, V. Oakley, and J. Navarro, "Investigations into the use of Laponite as a poulticing material in ceramics conservation," *Conservation Journal*, no. 22, 1997, Accessed: Aug. 13, 2023. [Online]. Available: http://www.vam.ac.uk/content/journals/conservation-journal/issue-22/investigations-into-the-use-of-laponite-as-a-poulticing-material-in-ceramics-conservation/
[13] J. Warda, I. Brückle, A. Bezúr, and D. Kushel, "Analysis of agarose, Carbopol, and Laponite gel poultices in paper conservation," *Journal of the American Institute for Conservation*, vol. 46, no. 3, pp. 263–279, 2007, doi: 10.1179/019713607806112260.
[14] Technical Services Branch for the USDA National Organic Program, "Attapulgite handling/processing," United States Department of Agriculture, Feb. 2010. Accessed: Aug. 10, 2023. [Online]. Available: https://www.ams.usda.gov/sites/default/files/media/Attapulgite%20TR.pdf
[15] S. Acebal and L. Vico, "Acid-base properties of aqueous suspensions of homoionic sepiolite and palygorskite," *Natural Resources*, vol. 8, no. 6, pp. 432–444, 2017, doi: 10.4236/NR.2017.86028
[16] Technical Services Branch for the USDA National Organic Program, "Attapulgite handling/processing," United States Department of Agriculture, Feb. 2010. Accessed: Aug. 10, 2023. [Online]. Available: https://www.ams.usda.gov/sites/default/files/media/Attapulgite%20TR.pdf
[17] Southern Clay Products, "Laponite RD Product Bulletin." [Online]. Available: https://conservationsupportsystems.com/system/assets/msds/LaponiteRD Data.pdf
[18] D. Stulik and R. Wolbers, "Project outcome, spin-offs, and future research needs," in *Solvent Gels for the Cleaning of Works of Art: The Residue Question*. Research in conservation, series. Los Angeles: Getty Publications, 2004.
[19] D. Stulik *et al.*, "*Solvent Gels for the Cleaning of Works of Art: The Residue Question,*" *Research in conservation, series*. Los Angeles: Getty Publications, 2004. Available: https://www.getty.edu/publications/resources/virtuallibrary/9780892367597.pdf
[20] J. Paul Getty Trust, "https://www.getty.edu/," Getty, 2023.
[21] R. Wolbers, "Terminology and Properties of Selected Gels," in *Gels in the Conservation of Art*, L. V. Angelova, B. Ormsby, J. Townsend, and R. Wolbers, Eds., London: Archetype, 2017, pp. 381–394.
[22] Lubrizol, "Neutralization procedures," manufacturer brochure bulletin 5, May 2011.
[23] Lubrizol, "Carbopol® polymer excipients - homopolymers, copolymers & interpolymers - Lubrizol," Lubrizol Pharmaceuticals. Accessed: Sep. 01, 2023. [Online]. Available: https://www.lubrizol.com/Health/Pharmaceuticals/Excipients/Carbopol-Polymer-Products
[24] N. Ravenel, "Pemulen® TR-2: an emulsifying agent with promise," *WAAC Newsletter*, vol. 32, no. 3, pp. 10–12, 2010.
[25] Lubrizol, "Introducing Pemulen® polymeric emulsifiers," Technical Data Sheet TDS-114, Oct. 2007. [Online]. Available: https://www.lubrizol.com/-/media/Lubrizol/Health/TDS/TDS-114_Introducing_Pemulen_Polymeric_Emulsifiers.pdf
[26] C. Stavroudis, "Gels: evolution in practice," in *Gels in the Conservation of Art*, L. V. Angelova, B. Ormsby, J. Townsend, and R. Wolbers, Eds., London: Archetype, 2017, pp. 209–217.
[27] Vanderbilt Minerals LLC, "Vanzan xantham gum," Brochure, Oct. 2018. [Online]. Available: https://www.vanderbiltminerals.com/resources/VANZAN_Brochure_English_Web.pdf

[28] G. Deckner, "Xanthan gum: a workhorse polysaccharide," Prospector Knowledge Center. Accessed: Feb. 06, 2024. [Online]. Available: https://www.ulprospector.com/knowledge/2368/pcc-xanthan-gum-workhorse-polysaccharide/
[29] I. Passeri, A. Finne, and A. Bezur, "Gel formulations coupled with mechanical cleaning techniques to remove old overpaint and pigmented coatings on a large altarpiece by Piero di Cosimo," in *Gels in the Conservation of Art*, L. V. Angelova, B. Ormsby, J. Townsend, and R. Wolbers, Eds., London: Archetype, 2017, pp 306–315.
[30] P. R. Varges, C. Costa, B. Fonseca, M. Naccache, and P. R. De Souza Mendes, "Rheological characterization of Carbopol® dispersions in water and in water/glycerol solutions," *Fluids*, vol. 4, no. 1, p. 3, 2019, doi: 10.3390/fluids4010003.
[31] P. Balcerzak, "Carbopol polymers for gels," Feb. 10, 2024.
[32] Conservation Support Systems, "Carbopol polymers," Conservation Support Systems. Accessed: Feb. 10, 2024. [Online]. Available: https://conservationsupportsystems.com/product/show/carbopol-polymers/gelling-mediums-additives
[33] C. Stavroudis, "Pemulen® revised: pHuck the pH meter," *WAAC Newsletter*, vol. 34, no. 2, p. 19, 2012.
[34] Lubrizol, "Polymer handling and storage," Manufacturer Brochure Bulletin 3, May 2011. [Online]. Available: https://www.lubrizol.com/-/media/Lubrizol/Health/Literature/Bulletin-03---Polymer-Handling-and-Storage.pdf
[35] R. Wolbers, *Cleaning Painted Surfaces: Aqueous Methods*. London: Archetype Publications, 2000.
[36] Lubrizol, "Dispersion techniques for Lubrizol pharmaceutical polymers," Bulletin 4, May 2011.
[37] D. Stulik and D. Miller, "Research into Solvent Gel Residues," in *Solvent Gels for the Cleaning of Works of Art: The Residue Questions*, Research in Conservation, series, Los Angeles: Getty Publications, 2004, pp. 19–83.
[38] S. Smith, "The removal of complex soiling on a collection of modern art bronzes using buffered pH-adjusted aquous gels," in *Gels in the Conservation of Art*, L. V. Angelova, B. Ormsby, J. Townsend, and R. Wolbers, Eds., London: Archetype, 2017, pp. 349–355.
[39] *Preparing a Pemulen gel from MCP and making an emulsion*, (Oct. 31, 2013). Accessed: Feb. 02, 2024. [Online Video]. Available: https://www.youtube.com/watch?v=2O5pYyc45Qo
[40] C. Stavroudis, "Using Pemulen with the MCP," *WAAC Newsletter*, vol. 32, no. 3, p. 16, 2010.
[41] N. Ravenel, "Pemulen TR2 / How to make the gel," PemulenTR2. Accessed: Feb. 08, 2024. [Online]. Available: http://pemulentr2.pbworks.com/w/page/15636418/How%20to%20make%20the%20gel
[42] K. deGhetaldi, B. Baade, Z. Voras, and E. Wroczynski, "Resurrecting a giant: using solvent gels and agueous systems to restore Villanova University's Triumph of David," in *Gels in the Conservation of Art*, L. V. Angelova, B. Ormsby, J. Townsend, and R. Wolbers, Eds., London: Archetype, 2017, pp. 157–164.
[43] KEGG, "Carbomer 934," Kyoto Encyclopedia of Genes and Genomes. Accessed: Feb. 10, 2024. [Online]. Available: https://www.genome.jp/dbget-bin/www_bget?dr:D03393#:~:text=Molecular%20weight%20is%20approximately%203%2C000%2C000.
[44] Lubrizol, "Carbopol 974P NF polymer," Product Specification. [Online]. Available: https://www.lubrizol.com/-/media/Lubrizol/Health/Literature/SPEC_Carbopol_974P_NF.pdf
[45] S. Crawford, "Carbopol polymers for gels," [Personal Communication]. Feb. 12, 2024.

[46] Lubrizol, "Carbopol 981 NF Polymer," Product Specification. [Online]. Available: https://www.lubrizol.com/-/media/Lubrizol/Health/Literature/SPEC_Carbopol_981_NF.pdf

[47] P. Mezzadri, G. Sidoti, and M. C. Gaetani, "The case of Capogrossi in Rome: trials and tests using gels for the cleaning of a contemporary wall painting," in *Gels in the Conservation of Art*, L. V. Angelova, B. Ormsby, J. Townsend, and R. Wolbers, Eds., London: Archetype, 2017, pp. 337–342.

[48] Lubrizol, "Carbopol® Ultrez 21 Polymer," Product Brochure, 2009. [Online]. Available: https://www.lubrizol.com/-/media/Lubrizol/Personal-Care/Documents/Brochures/Carbopol--Ultrez21-Polymer.pdf

[49] C. V. Horie, *Materials for Conservation: Organic Consolidants, Adhesives, and Coatings*, in Butterworths series in conservation and museology. London; Boston: Butterworths, 1987.

[50] Dow Chemical Company, "Methocel cellulose ethers," Technical Handbook, Sep. 2002. [Online]. Available: https://uolab.groups.et.byu.net/files/pipevisco/hints/methocel.pdf

[51] A. Sharma, S. Gautam, and S. Wadhawan, "Xanthomonas," in *Encyclopedia of Food Microbiology*, 3rd ed, C. A. Batt and M. L. Tortorello, Eds., Oxford: Academic Press, 2014, pp. 811–817. doi: 10.1016/B978-0-12-384730-0.00359-1.

[52] B. Aguilar Sánchez and R. Buentello Martínez, "Agarose gels with methyl ethyl ketone for cleaning a 19th-century document," in *Gels in the Conservation of Art*, L. V. Angelova, B. Ormsby, J. Townsend, and R. Wolbers, Eds., London: Archetype, 2017, pp. 107–109.

[53] S. Rapti, S. Boyatzis, S. Rivers, A. Velios, and A. Pournou, "Removing iron stains from wood and textile objects: assessing gelled siderophores as novel green chelators," in *Gels in the Conservation of Art*, L. V. Angelova, B. Ormsby, J. Townsend, and R. Wolbers, Eds., London: Archetype, 2017, pp. 343–348.

[54] L. L. Fiskovic, "Carbopol gel as a carrier for oxidizing agents in discoloured lead white conversion treatment," in *Gels in the Conservation of Art*, L. V. Angelova, B. Ormsby, J. Townsend, and R. Wolbers, Eds., London: Archetype, 2017, pp. 181–183.

[55] S. Kavda, S. Golfomitsou, and E. Richardson, "The effect of gelling agents and solvents on poly(methyl methacrylate) surfaces: a comparative study," in *Gels in the Conservation of Art*, L. V. Angelova, B. Ormsby, J. Townsend, and R. Wolbers, Eds., London: Archetype, 2017, pp. 331–336.

[56] E. Brown, "An unexpected surface: Research and treatment of a 19th century mounted oyster shell by Froment-Meurice," *Objects Specialty Group Postprints*, vol. 23, pp. 199–220, 2016.

[57] E. Guilminot, "The use of hydrogels in the treatment of metal cultural heritage objects," *Gels*, vol. 9, no. 3, p. 191, 2023, doi: 10.3390/gels9030191.

[58] E. Doherty, and S. Rivers, "The removal of lead-and oil-based overpaint from a plaster cast of Hermes Fastening his Sandal," in *Gels in the Conservation of Art*, L. V. Angelova, B. Ormsby, J. Townsend, and R. Wolbers, Eds., London: Archetype, 2017, pp. 122–126.

[59] A. Barberà Giné and S. Marín Ortega, "The removal of Paraloid-B72 coatings with aqueous gelled systems: Roman frescoes from Empúries, Catalonia," in *Gels in the Conservation of Art*, L. V. Angelova, B. Ormsby, J. Townsend, and R. Wolbers, Eds., London: Archetype, 2017, pp. 363–366.

[60] Lubrizol Noveon, "Neutralizing Carbopol®* and Pemulen™* polymers in aqueous and hydroalcoholic systems," Technical Data Sheet TDS-237, Sep. 2009. [Online]. Available: http://pemulentr2.pbworks.com/f/TDS-237_Neutralizing_Carbopol_Pemulen_in_Aqueous_Hydroalcoholic_Systems--PC.pdf

[61] P. Baglioni and D. Chelazzi, Eds., *Nanoscience for the Conservation of Works of Art*. Royal Society of Chemistry, 2013.
[62] H. Eriksson, I. Wedberg, J. Nessow, and M. Bronmark-Thorlund, "The use of Nanorestore gels in the conservation of lime-based wall paintings," in *Gels in the Conservation of Art*, L. V. Angelova, B. Ormsby, J. Townsend, and R. Wolbers, Eds., London: Archetype, 2017, pp. 270–273.
[63] M.-J. Jeong, A.-L. Dupont, and E. R. De La Rie, "Degradation of cellulose at the wet-dry interface: I—study of the depolymerization," *Cellulose*, vol. 19, no. 4, pp. 1135–1147, 2012, doi: 10.1007/s10570-012-9722-4.
[64] M.-J. Jeong, A.-L. Dupont, and E. R. De La Rie, "Degradation of cellulose at the wet–dry interface. II. Study of oxidation reactions and effect of antioxidants," *Carbohydrate Polymers*, vol. 101, pp. 671–683, 2014, doi: 10.1016/j.carbpol.2013.09.080.
[65] M. R. Sullivan, T. T. Duncan, B. H. Berrie, and R. G. Weiss, "Rigid polysaccharide gels for paper conservation: a residue study," in *Gels in the Conservation of Art*, London, UK: Archetype Publications Ltd., 2017, pp. 42–50.
[66] V. Stagno, A. Ciccola, R. Curini, P. Postorino, G. Favero, and S. Capuani, "Non-invasive assessment of pva-borax hydrogel effectiveness in removing metal corrosion products on stones by portable NMR," *Gels*, vol. 7, no. 4, p. 265, 2021, doi: 10.3390/gels7040265.
[67] U. Henniges, I. Brückle, H. Khaliliyan, and S. Böhmdorfer, "Gellan residues on paper: quantification and implication for paper conservation," *Heritage Science*, vol. 12, no. 1, p. 54, 2024, doi: 10.1186/s40494-024-01160-1.
[68] V. M. Gun'ko, I. N. Savina, and S. V. Mikhalovsky, "Properties of water bound in hydrogels," *Gels*, vol. 3, no. 4, p. 37, 2017, doi: 10.3390/gels3040037.
[69] P. Cremonesi and A. Casoli, "Thermo-reversible rigid agar hydrogels: their properties and action in cleaning," in *Gels in the Conservation of Art*, L. V. Angelova, B. Ormsby, J. Townsend, and R. Wolbers, Eds., London: Archetype, 2017, pp. 19–28.
[70] N. Pernodet, M. Maaloum, and B. Tinland, "Pore size of agarose gels by atomic force microscopy," *Electrophoresis*, vol. 18, no. 1, pp. 55–58, 1997, doi: 10.1002/elps.1150180111.
[71] F. Puoti *et al.*, "Evaluation of leather cleaning with a rigid hydrogel of gellan gum on two composite Amharic shields from the Museo Nazionale Preistorico Etnografico 'Luigi Pigorini', Rome," in *Gels in the Conservation of Art*, L. V. Angelova, B. Ormsby, J. Townsend, and R. Wolbers, Eds., London: Archetype, 2017, pp. 82–86.
[72] C. Delattre, S. Bouvet, and É. Le Bourg, "Gellan gum and agar compared to aqueous immersion for cleaning paper," in *Gels in the Conservation of Art*, L. V. Angelova, B. Ormsby, J. Townsend, and R. Wolbers, Eds., London: Archetype, 2017, pp. 57–61.
[73] Z. Miller, G. Whitby, and P. Garside, "Investigating the ability of phytate gel systems to treat iron gall ink at the British Library," in *Gels in the Conservation of Art*, L. V. Angelova, B. Ormsby, J. Townsend, and R. Wolbers, Eds., London: Archetype, 2017, pp. 77–81.
[74] M. Bertasa *et al.*, "An in-depth study on the agar gel effectiveness for built heritage cleaning," *Journal of Cultural Heritage*, vol. 47, pp. 12–20, 2021, doi: 10.1016/j.culher.2020.10.007
[75] S. Boral, A. Saxena, and H. B. Bohidar, "Syneresis in agar hydrogels," *International Journal of Biological Macromolecules*, vol. 46, no. 2, pp. 232–236, 2010, doi: 10.1016/j.ijbiomac.2009.12.008
[76] M. Arab, M. Yousefi, E. Khanniri, M. Azari, V. Ghasemzadeh-Mohammadi, and N. Mollakhalili-Meybodi, "A comprehensive review on yogurt syneresis: effect of processing conditions and added additives," *Journal of Food Science Technology*, vol. 60, no. 6, pp. 1656–1665, 2023, doi: 10.1007/s13197-022-05403-6

[77] Y. van Dyke, "Agarose-enzyme gels in paper conservation," in *Gels in the Conservation of Art*, L. V. Angelova, B. Ormsby, J. Townsend, and R. Wolbers, Eds., London: Archetype, 2017, pp. 101–106.

[78] S. Russick, N. Dobrowolski, B. Nosek, and R. Williams, "Gellan gum tips," *The Book & Paper Group Annual*, vol. 37, pp. 162–164, 2018.

[79] M. Cassanelli, I. Norton, and T. Mills, "Role of gellan gum microstructure in freeze drying and rehydration mechanisms," *Food Hydrocolloids*, vol. 75, pp. 51–61, 2018, doi: 10.1016/j.foodhyd.2017.09.013.

[80] V. Gold, Ed., *The IUPAC Compendium of Chemical Terminology: The Gold Book*, 4th ed. Research Triangle Park, NC: International Union of Pure and Applied Chemistry (IUPAC), 2019. doi: 10.1351/goldbook.

[81] M. Beaumont *et al.*, "Hydrogel-forming algae polysaccharides: from seaweed to biomedical applications," *Biomacromolecules*, vol. 22, no. 3, pp. 1027–1052, 2021, doi: 10.1021/acs.biomac.0c01406.

[82] M. Bertasa *et al.*, "A study of commercial agar gels as cleaning materials," in *Gels in the Conservation of Art*, London, UK: Archetype Publications Ltd., 2017, pp. 11–18.

[83] C. Magee, "High Acyl Gellan Gum for Parchment Conservation," *Book and Paper Annual*, vol. 38, pp. 112–119, 2019.

[84] Cinogel, "A good overview of scientific, technical and commercial aspects for gellan gum," Cinogel. Accessed: Feb. 23, 2024. [Online]. Available: https://www.cinogel.com/p/gellan-gum-overview.html

[85] C. S. F. Picone and R. L. Cunha, "Influence of pH on formation and properties of gellan gels," *Carbohydrate Polymers*, vol. 84, no. 1, pp. 662–668, 2011, doi: 10.1016/j.carbpol.2010.12.045.

[86] Q. Huang, Z. Liu, Y. Pei, J. Li, and B. Li, "Gelation behaviors of the konjac gum from different origins: A.guripingensis and A.rivirei," *Food Hydrocolloids*, vol. 111, p. 106152, 2021, doi: 10.1016/j.foodhyd.2020.106152.

[87] W. R. Thomas, "Konjac gum," in *Thickening and Gelling Agents for Food*, A. P. Imeson, Ed., Boston, MA: Springer US, 1997, pp. 169–179. doi: 10.1007/978-1-4615-2197-6_8.

[88] D. Yang *et al.*, "A review on konjac glucomannan gels: microstructure and application," *International Journal of Molecular Sciences*, vol. 18, no. 11, p. 2250, 2017, doi: 10.3390/ijms18112250.

[89] L. V. Angelova, P. Terech, I. Natali, L. Dei, E. Carretti, and R. G. Weiss, "Cosolvent gel-like materials from partially hydrolyzed poly(vinyl acetate)s and borax," *Langmuir*, vol. 27, no. 18, pp. 11671–11682, 2011, doi: 10.1021/la202179e.

[90] L. V. Angelova, B. H. Berrie, K. De Ghetaldi, A. Kerr, and R. G. Weiss, "Partially hydrolyzed poly(vinyl acetate)-borax-based gel-like materials for conservation of art: Characterization and applications," *Studies in Conservation*, vol. 60, no. 4, pp. 227–244, 2015, doi: 10.1179/2047058413Y.0000000112.

[91] I. Natali, E. Carretti, L. Angelova, P. Baglioni, R. G. Weiss, and L. Dei, "Structural and mechanical properties of 'peelable' organoaqueous dispersions with partially hydrolyzed poly(vinyl acetate)-borate networks: applications to cleaning painted surfaces," *Langmuir*, vol. 27, no. 21, pp. 13226–13235, 2011, doi: 10.1021/la2015786.

[92] S. Aruldass, V. Mathivanan, A. R. Mohamed, and C. T. Tye, "Factors affecting hydrolysis of polyvinyl acetate to polyvinyl alcohol," *Journal of Environmental Chemical Engineering*, vol. 7, no. 5, p. 103238, 2019, doi: 10.1016/j.jece.2019.103238.

[93] L. V. Angelova, B. Berrie, and R. Weiss, "Poly (vinyl alcohol)-borax 'gels': a flexible cleaning option," in *Gels in the Conservation of Art*, L. V. Angelova, B. Ormsby, J. Townsend, and R. Wolbers, Eds., London: Archetype, 2017, pp. 231–236.

[94] L. V. Angelova, B. Ormsby, J. Townsend, and R. Wolbers, Eds., *Gels in the Conservation of Art*. London: Archetype, 2017.

[95] PubChem, "Borax." Accessed: Feb. 24, 2024. [Online]. Available: https://pubchem.ncbi.nlm.nih.gov/compound/16211214

[96] "ICSC 0567 - sodium borate, decahydrate," UN International Labour Organization | International Chemical Safety Cards. Accessed: Feb. 24, 2024. [Online]. Available: https://www.ilo.org/dyn/icsc/showcard.display?p_lang=en&p_card_id=0567&p_version=2

[97] "Check your water quality | Help," Thames Water. Accessed: Feb. 23, 2024. [Online]. Available: https://www.thameswater.co.uk/help/water-quality/check-your-water-quality

[98] A. F. Maheux, "Cross-disciplinary uses for gellan gum in conservation," *The Book and Paper Group Annual*, vol. 34, pp. 69–79, 2015.

[99] Sigma, "Agarose," Product Information, 1996. [Online]. Available: https://www.sigmaaldrich.com/deepweb/assets/sigmaaldrich/product/documents/181/864/a9539pis.pdf

[100] D. Gulotta *et al.*, "Setup of a sustainable indoor cleaning methodology for the sculpted stone surfaces of the Duomo of Milan," *Heritage Science*, vol. 2, no. 1, 2014, doi: 10.1186/2050-7445-2-6.

[101] A. Giordano and P. Cremonesi, "New methods of applying rigid agar gels: from tiny to large-scale surface areas Ambra Giordano; Paolo Cremonesi," *Studies in Conservation*, vol. 66, no. 8, 2021, doi: 10.1080/00393630.2020.1848272

[102] A. Giordano, M. R. Caruso, and G. Lazzara, "New tool for sustainable treatments: agar spray—research and practice," *Heritage Science*, vol. 10, no. 1, p. 123, 2022, doi: 10.1186/s40494-022-00756-9

[103] M. Cushman, "Displayable colored gels?" Global Conservation Forum (ConsDistList). [Online]. Available: https://community.culturalheritage.org/discussion/displayable-colored-gels?ReturnUrl=%2fcommunities%2fcommunity-home%2fdigestviewer%3fListKey%3d757a8f16-505f-4323-8e74-e376757aa9f7%26CommunityKey%3dea3d002c-9fc3-4446-b7d2-c308f5faed13

[104] S. Barbisan, "Publication Contribution - Conservator's Cookbook," [Personal Communication] May 22, 2024.

[105] E. Al-Emam, A. G. Motawea, K. Janssens, and J. Caen, "Evaluation of polyvinyl alcohol–borax/agarose (PVA–B/AG) blend hydrogels for removal of deteriorated consolidants from ancient Egyptian wall paintings," *Heritage Science*, vol. 7, no. 1, p. 22, 2019, doi: 10.1186/s40494-019-0264-z

[106] C. Lee, F. Volpi, G. Fiocco, M. L. Weththimuni, M. Licchelli, and M. Malagodi, "Preliminary cleaning approach with alginate and konjac glucomannan polysaccharide gel for the surfaces of east Asian and western string musical instruments," *Materials*, vol. 15, no. 3, p. 1100, 2022, doi: 10.3390/ma15031100

[107] C. Lee *et al.*, "Biopolymer gels as a cleaning system for differently featured wooden surfaces," *Polymers*, vol. 15, no. 1, p. 36, 2022, doi: 10.3390/polym15010036

[108] G. Tobin and M. Sawicki, "Developing conservation practices for cleaning gilded surfaces: Applications for xPVOH-borax organogels to clean two gilded frames," in *Working Towards a Sustainable Past. ICOM-CC Preprints*, Valencia, Sep. 2023. [Online]. Available: https://www.icom-cc-publications-online.org/5600/Developing-conservation-practices-for-cleaning-gilded-surfaces--Applications-for-xPVOH-borax-organogels-to-clean-two-gilded-frames

[109] V. Prosapio, I. T. Norton, and E. Lopez-Quiroga, "Freeze-dried gellan gum gels as vitamin delivery systems: modelling the effect of pH on drying kinetics and vitamin release mechanisms," *Foods*, vol. 9, no. 3, p. 329, 2020, doi: 10.3390/foods9030329

[110] A. Shinsho *et al.*, "The thickening properties of native gellan gum are due to freeze drying–induced aggregation," *Food Hydrocolloids*, vol. 109, p. 105997, 2020, doi: 10.1016/j.foodhyd.2020.105997

[111] A. Nussinovitch, R. Velez-Silvestre, and M. Peleg, "Compressive characteristics of freeze-dried agar and alginate gel sponges," *Biotechnology Progress*, vol. 9, no. 1, pp. 101–104, 1993, doi: 10.1021/bp00019a015

[112] H. R. Moreira, L. P. da Silva, R. L. Reis, and A. P. Marques, "Tailoring gellan gum spongy-like hydrogels' microstructure by controlling freezing parameters," *Polymers (Basel)*, vol. 12, no. 2, p. 329, 2020, doi: 10.3390/polym12020329
[113] M. Cassanelli, V. Prosapio, I. Norton, and T. Mills, "Acidified/basified gellan gum gels: The role of the structure in drying/rehydration mechanisms," *Food Hydrocolloids*, vol. 82, pp. 346–354, 2018, doi: 10.1016/j.foodhyd.2018.04.024
[114] M. Zeece, "Chapter Seven - food additives," in *Introduction to the Chemistry of Food*, M. Zeece, Ed., Academic Press, 2020, pp. 251–311. doi: 10.1016/B978-0-12-809434-1.00007-4
[115] C. Rochas and M. Lahaye, "Average molecular weight and molecular weight distribution of agarose and agarose-type polysaccharides," *Carbohydrate Polymers*, vol. 10, no. 4, pp. 289–298, 1989, doi: 10.1016/0144-8617(89)90068-4
[116] A. M. Pegg, "8 - The application of natural hydrocolloids to foods and beverages," in *Natural Food Additives, Ingredients and Flavourings*, D. Baines and R. Seal, Eds., in Woodhead Publishing Series in Food Science, Technology and Nutrition. Woodhead Publishing, 2012, pp. 175–196. doi: 10.1533/9780857095725.1.175
[117] W. Wu *et al.*, "Effects of enzymatic konjac glucomannan hydrolysates on textural properties, microstructure, and water distribution of grass carp surimi gels," *Foods*, vol. 11, no. 5, p. 750, 2022, doi: 10.3390/foods11050750
[118] P. Baglioni *et al.*, "Micelle, microemulsions, and gels for the conservation of cultural heritage," *Advances in Colloid and Interface Science*, vol. 205, pp. 361–371, 2014, doi: 10.1016/j.cis.2013.09.008
[119] T. Giraud, A. Gomez, S. Lemoine, C. Pelé-Meziani, A. Raimon, and E. Guilminot, "Use of gels for the cleaning of archaeological metals. Case study of silver-plated copper alloy coins," *Journal of Cultural Heritage*, vol. 52, pp. 73–83, 2021, doi: 10.1016/j.culher.2021.08.014
[120] T. T. Duncan, B. H. Berrie, and R. G. Weiss, "A comparison between gel and swab cleaning: physical changes to delicate surfaces," in *Gels in the Conservation of Art*, L. V. Angelova, B. Ormsby, J. Townsend, and R. Wolbers, Eds., London: Archetype, 2017, pp. 250–256.
[121] E. Al-Emam, H. Soenen, J. Caen, and K. Janssens, "Characterization of polyvinyl alcohol-borax/agarose (PVA-B/AG) double network hydrogel utilized for the cleaning of works of art," *Heritage Science*, vol. 8, no. 1, 2020, doi: 10.1186/s40494-020-00447-3
[122] C. L. Scott, "The use of agar as a solvent gel in objects conservation," *Objects Specialty Group Postprints*, vol. 19, 2012, pp. 71–83.
[123] M. Barkovic, O. Diamond, and M. Cross, "The use of agar gel for treating water stains on an acrylic canvas," in *Gels in the Conservation of Art*, L. V. Angelova, B. Ormsby, J. Townsend, and R. Wolbers, Eds., London: Archetype, 2017, pp. 51–56.
[124] M. Brockman and E. Farek, "Investigation into reduction of foxing stains in paper," presented at the *Association of North American Graduate Programs in Conservation*, Queens University, Kingston, Ontario, 2018.
[125] C. Canevali *et al.*, "A multi-analytical approach for the study of copper stain removal by agar gels," *Microchemical Journal*, vol. 129, pp. 249–258, 2016, doi: 10.1016/j.microc.2016.07.007
[126] F. Gorel, "Assessment of agar gel loaded with micro-emulsion for the cleaning of porous surfaces," *CeROArt: conservation, exposition, restauration d'objets d'art*, 2010, doi: 10.4000/ceroart.1827
[127] N. Barbabietola *et al.*, "A safe microbe-based procedure for a gentle removal of aged animal glues from ancient paper," *International Biodeterioration & Biodegradation*, vol. 109, pp. 53–60, 2016, doi: 10.1016/J.IBIOD.2015.12.019

[128] R. Mazzoli, M. G. Giuffrida, and E. Pessione, "Back to the past—forever young: cutting-edge biochemical and microbiological tools for cultural heritage conservation," *Applied Microbiology and Biotechnology*, vol. 102, no. 16, pp. 6815–6825, 2018, doi: 10.1007/s00253-018-9121-3

[129] P. Bosch Roig, J. L. Regidor Ros, and R. Montes Estellés, "Biocleaning of nitrate alterations on wall paintings by Pseudomonas stutzeri," *International Biodeterioration & Biodegradation*, vol. 84, pp. 266–274, 2013, doi: 10.1016/j.ibiod.2012.09.009

[130] Y. van Dyke, "Practical Applications of Protease Enzymes in Paper Conservation," *The Book and Paper Group Annual*, vol. 23, pp. 93–107, 2004.

[131] M. Ferrari *et al.*, "Enzymatic laundry for old clothes: immobilized alpha-amylase from Bacillus sp. for the biocleaning of an ancient Coptic tunic," *Applied Microbiology and Biotechnology*, vol. 101, no. 18, pp. 7041–7052, 2017, doi: 10.1007/s00253-017-8437-8

[132] T. Markevičius, T. Syversen, E. Chan, N. Olsson, C. Skov Hilby, and R. Šimaitė, "Cold, warm, warmer: use of precision heat transfer in the optimization of hydrolytic enzyme and hydrogel cleaning systems," in *Gels in the Conservation of Art*, L. V. Angelova, B. Ormsby, J. Townsend, and R. Wolbers, Eds., London: Archetype, 2017, pp. 67–72.

[133] N. J. Borring and C. S. Hilby, "Enzymes in conservation: separating the Triumphal Arch by Albrecht Dürer with α-amylase embedded in agarose gels," in *IADA Congress XIII*, Berlin, 2015. doi: 10.13140/RG.2.1.1505.1609

[134] S. Sotgiu, S. Iannuccelli, S. Dominijanni, S. Puteo, L. Micheli, and C. Mazzuca, "Coupling gellan gels and electrochemical biosensors: real-time monitoring of cleaning and enzymatic treatments on a paper artwork," in *Gels in the Conservation of Art*, L. V. Angelova, B. Ormsby, J. Townsend, and R. Wolbers, Eds., London: Archetype, 2017, pp. 36–41.

[135] J. Vallieres, "Gellan gum: investigating applications as a solvent gel," Kingston (Ontario, Canada): Queen's University, 2013. [Online]. Available: https://www.queensu.ca/art/sites/artwww/files/uploaded_files/posterJaymeVallierescopy.pdf

[136] M. Tortora, F. Gherardi, E. Ferrari, and B. Colston, "Biocleaning of starch glues from textiles by means of α-amylase-based treatments," *Applied Microbiology and Biotechnology*, vol. 104, no. 12, pp. 5361–5370, 2020, doi: 10.1007/s00253-020-10625-9

[137] G. Petrella *et al.*, "A new sustainable and innovative work for paper artworks cleaning process: gellan hydrogel combined with hydrolytic enzymes," *International Journal of Conservation Science*, vol. 7, no. speciall, 2016.

[138] C. Mazzuca *et al.*, "Cleaning of paper artworks: Development of an efficient gel-based material able to remove starch paste," *ACS Applied Materials and Interfaces*, vol. 6, no. 19, pp. 16519–16528, 2014, doi: 10.1021/am504295n

[139] F. Palla *et al.*, "Cold-active molecules for a sustainable preservation and restoration of historic-artistic manufacts.," *International Journal of Conservation Science*, vol. 7, no. SI1, pp. 239–246, 2016.

# 4 Cleaning Agents

### Caution

Before we go any further, despite its inclusion in the 'recipes', with most cleaning systems there is little finesse to their preparation, but much nuanced understanding required for their use. This short section is intended to briefly summarise some of the key concepts and mechanisms for those with some degree of pre-existing knowledge.

Cleaning is a *highly* interventive approach, potentially swelling paints, destabilising colourants, dyes, evidence of use, sizing, and other desirable and undesirable materials. This should be attempted only with the utmost caution and extensive testing prior to use. See 'Health and Safety' in Chapter 1 for information on best practices when working with hazardous or reactive chemicals.

Interested parties are strongly encouraged to see some of the books, articles, and online lectures listed in the bibliography. Those looking to employ carefully designed bespoke surface cleaning systems are even more strongly encouraged to see Stavroudis' work with the Modular Cleaning Program (MPC).

## SURFACTANTS

Consultant: Joanne Hackett

### *Material Information*

Precision Required: Medium

#### *Introduction*

'Like dissolves like' is perhaps the most concise aphorism in solutions chemistry, reflecting the propensity for non-polar, oily-feeling compounds to dissolve into non-polar solvents and watery-feeling polar compounds into polar solvents like water. However, one major challenge in cleaning is that of the reverse, dissolving unlike into unlike. One method for this is found in the idea of detergency, where soiling is brought into solution with the assistance of an

DOI: 10.4324/9781003391708-5

intermediary compound known as a surface-active agent, or surfactant. These molecules exhibit a dual character which enables interactions with both solvent types, with surfactants essentially escorting one compound into another [1, 2].

Surfactants are typically semi-linear amphiphilic organic molecules which lower the surface tension of aqueous systems and, in the right conditions, will form micelles in solution. There are four main types of surfactant, each described by the electric nature of the 'head': anionic, non-ionic, cationic, and amphoteric [1, 2]. Of these, only the first two will be discussed in any detail as these are the types that are routinely used in conservation cleaning [3, 4].

*Composition*

Surfactant formulation and production is a multi-billion-dollar industry, with formulations popping in and out of production in very short time scales. While consistency is always an issue in sourcing conservation materials, surfactant formulation is more closely guarded and quickly changing than most. As a result, while some mention will be made of specific proprietary surfactants, composition will be discussed only in general terms. In the 'Technical Information' section every effort will be made to provide chemical information where possible, but this information is frankly often not freely available.

Surfactants are typically composed of a long tail with a hydrophilic head. The character of the head is what determines type. A non-ionic surfactant has a large polar head, but it does not dissociate in solution. The head of ionic surfactants comprises a simple salt, and upon exposure to polar solvent, the head can separate (or dissociate) into its ionic parts. One-half of the pair freely floats off into solution, leaving either a positively charged (cationic) or a negatively charged (anionic) head attached to the long tail [1]. Of these, only polar and anionic surfactants are typically used in conservation activity [3, 4]. Cations (surfactant tails with positively charged heads) are attracted to aged decorative or varnished surfaces and have a tendency to bond with fibres in a manner similar to dyes, making them very difficult to remove [5]. The final class of surfactants is amphoteric (Zwitterionic), which can show either negative or positive charge within the same molecule, but are very expensive and not attested within the conservation literature [4].

*Classification and Types*

As mentioned, anionic surfactants are characterised by a small 'head' which ionises in water to produce a free cation and a negatively charged anion at the head of a very long tail. Generally speaking, these can provoke more efficient soil removal, but are less stable and more difficult to wash out, requiring warmer water to solubilise [4]. Anionics also work best in slightly alkaline conditions and are therefore often used in the cleaning of cellulose fibres.

Non-ionic surfactants are structurally quite distinct from anionics. As the same suggests, these do not ionise in solution, but instead consist of a large polar head and a short nonpolar tail. These are more chemically stable, tolerating both the presence of electrolytes as well as small changes in pH. Non-ionic surfactants tend to work most efficiently in slightly acidic conditions and are therefore better suited for cleaning acidic surfaces or proteinaceous fibres [4].

Broadly speaking, at comparable hydrophobic chain lengths, non-ionic are more surface active than anionic and are also quicker to form micelles. Although comparatively small, the negatively charged heads of anionic surfactants repel each other, hindering compact micelle formation. This is reflected in critical micelle concentration (CMC) trends for the two types, with non-ionic detergents generally exhibiting lower CMC (0.005–0.05%) than anionic (0.05–0.5%) [1] (Figures 4.1–4.3).

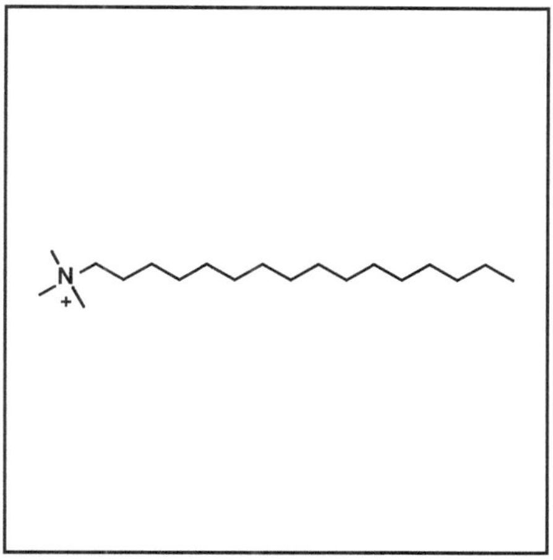


*Figure 4.1* Structure of cetrimonium (bromide), a cationic surfactant. Cationic surfactants are not typically used in conservation; their inclusion is illustrative of structure.

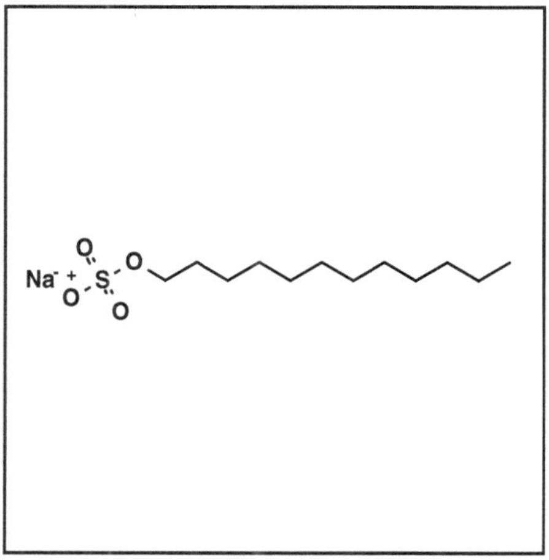


*Figure 4.2* Structure of sodium dodecyl sulfate, an anionic surfactant.

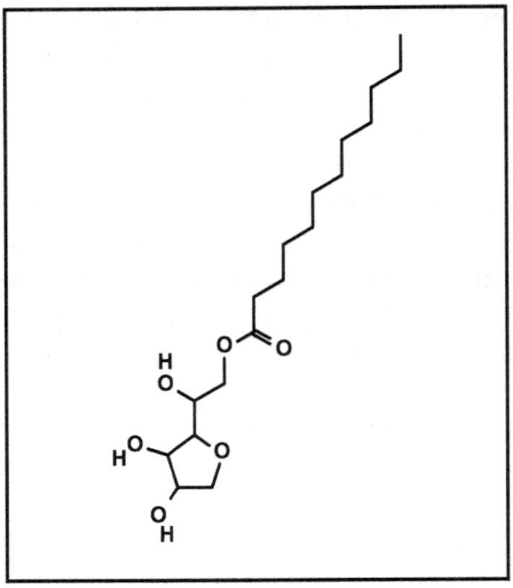

*Figure 4.3* Structure of sodium monolaurate, a non-ionic surfactant.

*Behaviours and Characteristics*

WETTING

As discussed, surfactant molecules contain both a charged/polar, "hydrophilic" head and a non-polar "hydrophobic" tail. This dual character (amphiphilia) allows each end of the molecule to exhibit different electrochemical attractions. The head of the surfactant will therefore be attracted to the dipole of water, causing the surfactant to orient itself with the head closer to the water and the tail closer to any non-polar surface nearby (including the air above).

As surfactants accumulate on the surface of the water, this disrupts the hydrogen bonding that gives water its strong cohesion. This therefore lowers the surface tension of the solution allowing for better wetting and flow. Essentially, this allows the water to stop sticking so closely to itself, instead spreading out more rapidly and widely[1, 6].

MICELLES AND CMC

As more surfactant is added to a solution and the air-water interface becomes crowded with surfactant molecules, the molecules will begin to aggregate into little, typically spherical 'pockets' called micelle spheres. In aqueous solution these aggregates form with the hydrophobic tails clustered together in the middle and the hydrophilic heads on the outside, closest to the water.

The key to detergency is found in the ability of micelles to form around other water-insoluble materials, 'rolling up' the undesired materials into a sphere whose outermost crust is water soluble. This then allows the unlike material to be brought into the unlike solution.

The beginning of micelle formation is marked by a sudden upswing in solubility as the change in spatial orientation encourages further dissolution. Micelles won't form if a solution is too dilute and the ratio of surfactant to solvent too low. The concentration at which micelles suddenly begin to aggregate is known as the critical micelle concentration (CMC) [2].

This key point is specific to each surfactant, and as cleaning requires micelle formation and micelles only occur above the CMC, the concentration of surfactant in solution must exceed the CMC for cleaning to occur. Thankfully, CMC is vital information for formulation chemists; it is conveniently often available in manufacturer-provided technical literature.

### DETERGENCY

For detergency to occur, there must be a sufficient concentration of surfactant to both allow for micelle formation and to ensure that the supply is not immediately exhausted. However, for conservation purposes, more is not always better, and the greater the concentration of surfactant, the higher the number of rinses required to fully remove it.

A good balance must therefore be struck between having enough surfactant to support detergency and resist depletion, while also minimising the risk of deposition and the impact of repeated rinsing. Literature from textile and paintings conservation indicates that the sweet spot is often found between two and five times the CMC for most surfactants [3, 4].

While the CMC can be determined from technical literature, conditions within a solvent system can and will affect this figure. Variables such as the particular molecular weight, shape, and polarity of the surfactant as well as the composition, concentration, pH, and temperature of the solution can all shift the CMC. This is all to say that while understanding and calculating the desired CMC for a given system is an important step, in practical application, some amount of inconsistency is to be expected. This again, and always, is why testing is vital prior to undertaking any wet cleaning process.

### HYDROPHILE-LIPOPHILE BALANCE (HLB) NUMBER

The HLB number is intended to describe the ability of non-ionic surfactants to form emulsions. Low numbers reflect molecules with either strong hydrophilic *or* strong lipophilic tendencies, while high numbers have both hydro and lipophilic tendencies within the same molecule [3]. This is intended to be predictive, with high numbers (10 and up) building more stable 'oil into water' emulsion systems and lower numbers (<10) separating out of aqueous systems more swiftly but forming stable 'water into oil' emulsions [6]. There are several considerable downsides to working with HLB, not least of which is the fact that it treats surfactants in isolation, when micelle formulation is highly contingent on solution conditions (pH, temperature, etc.). This will gain greater significance in the section on emulsions in Chapter 3 but is included here for context.

*Temperature Effects*

The solubility of a surfactant and the changes this undergoes is dependent on both concentration and temperature. The point at which solubility suddenly increases as related to the concentration of the solution is called the CMC. When the solubility of a surfactant changes as a function of temperature, this is called either the Cloud point or the Krafft point.

CLOUD POINT

When heated above a certain temperature, solutions of non-ionic surfactants tend to become more opaque and white. As the solution warms, non-ionic surfactants exhibit decreased bonding with neighbouring water molecules which results in conformational changes within the surfactant shape and reorganisation of the micelle structure. Briefly, the water bonded to the ether linkages on the long tails separate away and the tail stretches out into a more linear conformation [1]. This reduces both the hydrophilia of the tail and the steric hindrance between the surfactant molecules and allows for tighter grouping within micelles. These larger micelle aggregations and the increasingly organised relationship between them causes the solution to become cloudy white [1]. Importantly for conservators, this reduced hydrophilic affinity of the straightened tail corresponds with a reduced solubility of the surfactant within solution.

Essentially, the solubility of a non-ionic surfactant will decrease when it is warmed above a certain temperature. The point at which this takes place is called the cloud point because larger, more crystalline micelles and the undissolved surfactants will cause the appearance of the solution to become cloudy white. As a result, non-ionic surfactants are typically more soluble at lower temperatures (below 30°C for many).

KRAFFT POINT

Ionic surfactants are 'activated' by the dissociation of the salt which forms the head, with one charged ion entering solution and the other remaining covalently bonded to the long tail. Ionic dissociation is an equilibrium type reaction and driven by kinetics. As such, increases in temperature will directly correspond to faster rates of dissociation and therefore greater solubility. As you would expect with most ionic solutes, the solubility of ionic surfactants increases along with the temperature.

However, this is not a linear progression, and the Krafft point marks the temperature at which the solubility of an ionic detergent suddenly *increases*. Although distinct from (and in some ways in inverse of) the cloud point, the Krafft point represents a change in solubility resulting from changes in spatial orientation. In a solution with a concentration over the CMC, but a temperature below the Krafft point, the surfactant molecules will arrange in a bilayer structure which forms a hydrated solid. Above the CMC and the Krafft point, the surfactant molecules will 'roll up' into micelles, entering solution allowing further dissolution [1].

*Key Ideas*

Due to the complexity of the topic, the 'summary' given above is a bit bulkier than this book typically aims for. Please see the summary of the summary below for gross over-generalisations and super-reductive statements

- Surfactants used in conservation are typically anionic or non-ionic.
- Detergency can require the aggregation of micelles, and micelle formation requires a minimum concentration (CMC) and specific temperature ranges (cloud and Krafft points).
- Anionics will cloud and become less water soluble when cool (below 40°C).
- Non-ionics will cloud and become less water soluble when warm (above 30°C).

### ***Preparation: Surfactants***

*Techniques and Tips*

CALCULATING QUANTITY

Once you determine which detergent to use, you'll need to figure out how much. A table of CMC data has been compiled in the 'Technical Information' section; however, these figures are prone to change without notice as manufacturers update formulations. It's therefore valuable to determine practical CMC figures from manufacturer supplied data. To do so a simple unit conversion is easiest. A longer example is given in Chapter 1, but a short summary is also included below.

Values from technical data sheet (TDS) provided by Dow
Ecosurf EH-3 CMC is given as 480 ppm
Density 0.9651 g/mL = 965.1 kg/m$^3$

$$\frac{\text{PPM} \times \text{density}\left(\frac{\text{kg}}{\text{m}^3}\right)}{1000} = \text{mg/L}$$

$$\frac{480 \times 965.1\ \text{kg/m}^3}{1000} = 463.248\ \text{mg/L}$$

$$463.248\ \text{mg/L} = 0.463\ \text{g/L}$$

TEMPERATURE

Just a quick reminder that Cloud/Krafft points will influence detergency and solubility. Broadly speaking, anionics are more soluble at warmer temperatures, and non-ionics at cooler ones. It might be beneficial to place a small amount of water in a refrigerator or microwave to encourage the initial dissolution.

MIXING

Particularly in immersion cleaning, undesired soiling can be lifted away during detergency but then redeposited back on the object's surface. To limit this redeposition and help keep the soiling suspended in solution, textile conservators often add a small amount of sodium carboxymethyl cellulose. However, as discussed in Chapter 2 on cellulose ethers, this material can be difficult to dissolve. It is therefore recommended to fully dissolve this separately before adding it to a wash bath or cleaning solution.

APPLICATION METHODS

Surfactants can be applied in numerous ways, a few of which are discussed in greater detail elsewhere. They can be used in immersion baths for textiles, soaked into sponges or cotton swabs, and dabbed or rolled onto surfaces, or added to gels and poultices. In fact, pretty much every aqueous cleaning system in conservation supports the addition of surfactants.

REMOVAL/CLEARANCE

No matter the application method chosen, upon completion of the cleaning process, the surfactant must be fully rinsed away. This is particularly true of ionic surfactants which have been shown to linger on the surface and have the potential to alter the pH of a surface [7]. The amount of rinsing required will depend on the type of surfactant, its concentration, the application method, and the nature of the substrate. Two rinses in water is considered sufficient for many applications, but this should be considered a minimum if there is any uncertainty. Rinses in this context can refer to immersions, rolled swabs, gel applications, or other approaches and typically mirror the application methods.

*Materials*

Volumetric measure: flask, graduated cylinder, measuring cups, etc.
If supplied as solid: Analytical balance and weighing boats
If supplied as liquid: pipette
Stirring implement, glass preferrable

*Instructions*

1. Calculate desired amount of surfactant (see CMC and dilution in 'Calculations' section of Chapter 1).
2. Decant surfactant using balance or pipette.
3. Add the surfactant to the larger volumetric measure and top up to roughly half of the final volume with warm or cool water.
4. Add any additives, such as chelation agents or buffers.
5. Add the remaining water, measure, and adjust pH as required.
6. Stir gently or swirl within container until all components are fully dissolved.

*Clean Up*

Many, many surfactants and/or their degradation products are dangerous to aquatic life and chemical safety information should be checked prior to use. If using small volumes and high concentrations, water can be allowed to evaporate from the container, and the dried or liquid waste can be sealed and binned or placed with chemical waste. Never release directly into roadside drains, storm sewers, or onto the ground.

It's true, as some have pointed out, that the 'dangerous to aquatic life' designation and accompanying dead fish symbol can be easy to misinterpret. These symbols are given to materials which are found to have negative ecological effects when discharged directly into watercourses. It does not necessarily take into account the effect of wastewater processing, of which surfactant clearance is a key part [8].

However, this isn't to say that conservators are fine to simply wash excess surfactant down the sink rather than tipping it down outdoor drainage. Although modern waste processing is designed to remove surfactants, much like our own, that industry is also grappling with the breakneck pace at which contemporary surfactants are being developed and employed. The variety and staggering volume of surfactants in contemporary products means that their removal is a complex and evolving science [9, 10].

***Technical Data***

Technical data sheets (TDS) will usually contain information on CMC, HLB, and cloud point. These are often freely available on suppliers' or manufacturers' websites. However, there are different ways of determining these descriptors and the value supplied by the manufacturer can differ from experimental data. These values should therefore be used as a starting point rather than a firm rule (Table 4.1).

*Table 4.1* Collection of some surfactants, technical information, and examples of use

| *Name* | | *Composition* | *Type* | *Solubility* | *HLB* | *CMC (Ppm 25 °C)* | *Density (g/mL)* | *Cloud point (°C)* | *Use* | *Cite* |
|---|---|---|---|---|---|---|---|---|---|---|
| Aerosol OT-75 | Solvay | Sodium Dioctyl Sulfosuccinate | Anionic | Water, more unlisted | - | - | | - | Provided as 75% solid in a mixture of ethanol and water<br>Attested as wetting agent, emulsifier (1%) and in degreasing of aluminium surfaces (10%) | [11, 12] |
| APG 12-14 | Various | Laurel glucoside | Non-ionic | Water, alcohol | 12 | ~3% | 1.16 | | Emulsifier. Use in conservation not attested | [13] |
| Brij-35 | Croda | Dodecan-1-ol, ethoxylated | Non-ionic | Water, alcohol | 16.9 | 11 | 1.03 at 20 °C | >100 | Painted surfaces, emulsifier | [14] |
| Brij S100 | Croda | Polyoxyethylene (100) stearyl ether | Non-ionic | Water, alcohol | 18.8 | - | 1.1 at 20 °C | 100 | Replacement for Brij 700, but not currently attested in literature | [15] |
| Dehypon LS45 | BASF | Fatty alcohol alkoxylate | Non-ionic | Water, alcohols | | 50 (F) | | 20 | Textile cleaning. Low cloud point can make it difficult to use in summer months. | [8, 16–18] |
| Dehypon LS54 | BASF | Fatty alcohol alkoxylate made from a C12-C14- fatty alcohol (1 mole) ethylene oxide (5 moles) and propylene oxide (4 moles) | Non-ionic | Water, alcohols, aromatic hydrocarbons | - | 15 | 0.97 at 20 °C | 30 1% | Textile cleaning.<br>Replacement for Synperonic N, but with higher cloud point than Dehypon LS45.<br>Research indicates that it is most effective at 7.5x the CMC (0.3% w/v). | [8, 18, 19] |

| | | | | | | | | | | |
|---|---|---|---|---|---|---|---|---|---|---|
| Ecosurf EH-3 | Dow | Ethoxylated or propoxylated branched alcohol (2-ethyl hexanol) | Non-ionic | Water, polar organic and chlorinated solvents. | 7.9 | 480 | 0.9651 at 40 °C | 52 10% | Microemulsions for acrylic paintings, textile cleaning<br>DOW states that they are readily biodegradable and have an aquatic toxicity EC50> 10 mg/L. | [20, 21] |
| Ecosurf EH-6 | Dow | Ethoxylated or propoxylated branched alcohol (2-ethyl hexanol) | Non-ionic | Water, polar organic and chlorinated solvents. | 10.8 | 914 | 0.9897 at 40 °C | 43 10% | Microemulsions for acrylic paintings, textile cleaning<br>DOW states that they are readily biodegradable and have an aquatic toxicity EC50> 10 mg/L. | [21, 22] |
| Hostapon TPHC | Clariant | Oleic acid methyl tauride sodium salt | Anionic | water | - | 43 | | - | Used for textile immersion cleaning predominantly in the UK. Designed as a foam booster for shampoo.<br>Cloud point unlisted, but has been found routinely effective in Glasgow in the winter.<br>Note: TDS gives 28% NaCl | [16, 23] |
| Span 20 | Croda | Sorbitan monolaurate | Non-ionic | Xylene, mineral oil, other non-polar. **INsoluble** in water | 8.6 | - | 1.058 | - | Emulsifier. Use in conservation not attested | [24] |

(*Continued*)

*Table 4.1* (Continued)

| *Name* | | *Composition* | *Type* | *Solubility* | *HLB* | *CMC (Ppm 25 °C)* | *Density (g/mL)* | *Cloud point (°C)* | *Use* | *Cite* |
|---|---|---|---|---|---|---|---|---|---|---|
| Tergitol 15-S-3 | Dow | secondary alcohol ethoxylate | Non-ionic | Polar organic and chlorinated solvents.<br>Insoluble in water | 8.0 | - | 0.918 at 20 °C | - | Cleaning phonograph and vinyl records.<br>2.5 mL Tergitol 15-S-3 and 2.5 mL Tergitol 15-S-9 per 1 L distilled water | [25, 26] |
| Tergitol 15-S-9 | Dow | secondary alcohol ethoxylate | Non-ionic | Water, polar organic and chlorinated solvents. | 13.3 | 52 | 1.006 at 20 °C | 60 1% | Cleaning phonograph and vinyl records.<br>2.5 mL Tergitol 15-S-3 and 2.5 mL Tergitol 15-S-9 per 1 L distilled water | [26, 27] |
| Triton X-100 | Dow | Octylphenol Ethoxylate | Non-ionic | Water, alcohols, organic and chlorinated solvents | 13.4 | 189 | 1.061 at 20 °C | 66 1% wt | **No longer in use**.<br>Very popular before mid-1990s<br>Although not harmful on its own, one of its decomposition products is an oestrogen mimic known to damage fish reproductive systems. | [10, 28] |
| Tween 20 | Croda | Polyethylene glycol sorbitan monolaurate | Non-ionic | Water, polar solvents likely | 16.7 | 7 | 1.095 at 25 °C | >70 | Painted surface, emulsifier. | [3, 29] |
| Tween 80 | Croda | Polyethylene glycol sorbitan monooleate | Non-ionic | Water, polar solvents likely | 15.0 | 2.5 | 1.064 (no T) | 90 | Painted surface, emulsifier. | [3, 30–32] |
| WA paste | Orvus | Sodium lauryl sulphate | Anionic | Hot and cold water | - | 2900 | | - | Used for textile immersion cleaning predominantly in the US. High foaming (2%) | [8] |

## BUFFERS AND PH ADJUSTMENT

Consultants: Joyce Townsend and Thomas Higgins

### *Material Information*

Precision Required: Medium

### *Overview*

As with temperature and concentration, small changes to the pH of a solution can strongly affect the chemical behaviour within it. In the context of conservation, some solutions require specific conditions to work effectively over the short term, and these can usually be modified with simple pH adjustment. Buffers, on the other hand, are used for systems that require a very specific pH to be maintained over long periods of time and against changing conditions within the solution.

A short health and safety note before further discussion: many of these materials are harmful or toxic to human or aquatic life. Read all SDS information prior to use, ensure you maintain good lab etiquette, and wear appropriate personal protective equipment (PPE) such as gloves and eye protection. Hazardous materials should be disposed of as per the specific guidance for each type. See the section on health and safety in the first chapter of this text.

### *Composition, Behaviours, and Characteristics*

As a quick reminder, the pH scale records the relative concentration of $H_3O^+/H^+$ to OH- ions in an aqueous solution measured in moles per litre. Although technically unlimited, it is summarised to run from 1 (very acidic) to 14 (very alkaline), with pH 7 as the neutral middle point. It is an inverse logarithmic scale, with each single step up or down the ladder reflecting a change in 10 times the actual concentration of ions in solution. This means that a solution of pH 2 is actually ten times more acidic than a solution of pH 3. Adding more $H_3O^+$ (hydronium ions, sometimes also just called protons $H^+$) to a solution will drive the acidity up and the pH down, while the addition of OH- (hydroxide ions) will result in a more alkaline, higher pH solution.

A buffer is a weak conjugate acid-base pair that will maintain a specific pH level in solution despite the addition of other acids or bases. They are particularly helpful in circumstances where reactions occurring within the solution and during the cleaning process, can drive the pH out of the desired range. This is particularly important in the use of chelation agents and enzymes.

When choosing a buffer, the general advice is to select one which has an acid dissociation (pKa) within 1 unit of the desired **pH** range for the system [5]. It is important to select the right pH adjustment for the material and the buffer system desired. Some common combinations have been collated in the 'Technical Information' section, but suppliers' websites will often specify the ideal buffering system for compounds whose behaviour is highly pH dependent.

***Preparation: pH Altering Solutions***

*Techniques and Tips*

Think about what might be released into solution when selecting any buffers or chemicals for pH adjustment. For example, chloride-containing compounds should be avoided when working with metals, as chloride ions are known to facilitate degradation of many heritage metals [5]. Finally, solutions adjusted with significant volumes of sodium hydroxide have been recorded as unsuitable for cleaning gypsum, as they darken the surface [33].

CALCULATING MOLAR CONCENTRATION

For ease of calculation and practicalities of use, most pH adjustments in conservation are completed with solutions of 1 M to 10 M concentration acids and bases. A quick reminder that molarity (M) = moles solute/litres of solvent or M = n/V. See 'Calculations' for a (slightly) more thorough discussion of molarity and several examples, but a rough outline is given below.

Converting molar concentration/molarity (M) to grams per litre.

1. Determine the formula of solute (in this instance, likely an acid or base).
2. Calculate the mass of one mole of solute using the formula weight. The formula weight is the sum of the atomic weights of the correct number of each element in the formula. Sodium hydroxide (NaOH) contains one each of sodium, oxygen, and hydrogen.
   1 sodium + 1 oxygen + 1 hydrogen
   22.99 + 16.00 + 1.01 = 40.00 g/mol of NaOH
3. Convert from desired final concentration (1 M here) to grams using molar mass as a conversion factor.

$$1\,\mathrm{M} = \frac{1\,\mathrm{mol}}{\mathrm{L}}$$

$$\frac{1\,\cancel{\mathrm{mol}}}{\mathrm{L}} \times \frac{40\,\mathrm{g}}{\cancel{\mathrm{mol}}} = 40\,\mathrm{g/L}$$

4. Mix: fill a 1L volumetric container two-thirds with water, slowly add the base, and once dissolved, then add the remaining water. Again, ensure that you're exercising caution when diluting caustic materials. Dissolution of salts like sodium hydroxide liberates a lot of heat and if done too quickly can cause splatter of the solution and lead to chemical burns. The heat can also damage or crack non-thermal glassware.

DILUTING STRONG ACIDS AND BASES

Strong acids and bases are often stored in concentrated forms which require dilution prior to use. There are several reasons for this, not least of which is that these solutions can have a very long shelf life If stored appropriately (cool and dark).

When diluting acids always add the acid to the water, and never the other way around. When acid mixes with water, it can release a large amount of heat very quickly. With concentrated acids, incorrect mixing can lead to eruptive boiling, splattering, and burns. Please remember to wear correct PPE and always 'do as you oughta and add acid to water'. Goofy, yes, but all the best mnemonics seem to be.

Note that with dilution, it doesn't matter what specific units are used, so long as it's consistent. It can be moles per litre, percent, or ppm, so long as the units match on both sides of the equation. Further examples discussed in Chapter 1, but the basic equation is given here.

$$M_1 V_1 = M_2 V_2$$

CALCULATING BUFFER QUANTITY

Once you determine which buffer you would like to use, you will need to calculate how much is required. To do so you can either use the reorganisation of the Henderson–Hasselbalch equation discussed in Chapter 1 or rely on online tools like the 'Buffer Calculator' operated by Sigma-Aldrich Millipore [34], discussed with examples in 'Calculations'.

*Materials*

PPE: see chemical-specific safety guidance and the information in 'Health and Safety'
Volumetric measure: flask, graduated cylinder
Analytical balance and weighing boats
Stirring implement, glass preferrable

*Instructions*

1. Calculate the amount of solute needed. Measure out and set aside.
2. Fill the volumetric measure with roughly 80% the final volume of water.
3. Slowly add some of the acid or base to the volumetric container and mix well.
4. Check the temperature of the container with the side of your gloved hand. This may be warm, but ensure that it isn't hot before proceeding.
5. Slowly add the rest of the rest of the acid or base mixing regularly and keeping an eye on the temperature.
6. Add the rest of the remaining water, until the desired final volume is met.

***Technical Data***

*Table 4.2* Names and CAS are for one-half of the weak acid/base pair that make up the buffer

| *Buffer* CAS number(s) | *pKa* | *pH range* | *pH adjustment* CAS number |
|---|---|---|---|
| Citric acid<br>*77-92-9* | 3.13<br>4.76<br>6.40 | 2.2–6.5<br>3.0–6.2<br>5.5–7.2 | Sodium hydroxide *1310-73-2*<br>Disodium phosphate *7558-79-4*<br>Sodium citrate *994-36-5* |
| Formic acid<br>*64-18-6* | 3.75 | 2.8–4.6 | Hydrochloric acid *7647-01-0* |
| Acetic acid<br>*64-19-7* | 4.77 | 3.6–5.6 | Sodium acetate *127-09-3* |
| Monopotassium phosphate<br>*7778-77-0* | 2.15<br>6.82<br>12.38 | 5.0–8.0<br>5.8–8.0 | Disodium phosphate *7558-79-4*<br>Sodium hydroxide *1310-73-2* |
| Monosodium phosphate<br>*7558-80-7* | 2.15<br>6.82<br>12.38 | 5.8–8.0 | Disodium phosphate *7558-79-4*<br>Citric acid *77-92-9* |
| (MES) 2-(N-Morpholino) ethanesulfonic acid<br>4432-31-9 | 6.15 | 5.2–7.1 | Sodium hydroxide *1310-73-2* |
| Triethanolamine (TEA)<br>*102-71-6* | 7.76 | 7.0–8.3 | Hydrochloric acid *7647-01-0* |
| Tris(hydroxymethyl) aminomethane<br>*77-86-1* | 8.08 | 7.2–9.0 | Hydrochloric acid *7647-01-0* |
| Diethanolamine<br>*111-42-2* | 8.8 | 7.8–9.8 | Hydrochloric acid *7647-01-0* |
| Ammonium hydroxide<br>*1336-21-6* | 9.24 | 8.2–10.2 | Ammonium chloride *12125-02-9* |
| Sodium Carbonate<br>*497-19-8* | 10.33 | 9.2–10.6 | Sodium hydrogen carbonate<br>*144-55-8* |

Multiple pKas are provided for compounds with more than one acidic site. Data collated from Rivers and Umney, Wolbers, and Lange and Dean [3, 5, 35].

## Chelation Agents

Consultants: Joyce Townsend and Thomas Higgins

***Material Information***

Precision Required: High

*Overview*

Chelating agents are molecules with two or more sites that donate electron pairs (sometimes called dentae or teeth) that will bond with metal ions to form a coordinated, soluble structure. While chemically and functionally completely distinct from surfactants, they can serve an analogous function, acting as an intermediary and coordinating with an insoluble material in a manner which makes it soluble [36].

Chelates change the properties of an ion by either making it soluble in various aqueous and non-aqueous solutions or by trapping it, thereby prohibiting participation in undesirable reactions [37].

Since they were first adopted by conservators in the mid-twentieth century, chelating agents (sometimes also referred to as sequestering agents) have been used on a broad range of materials from stone to plaster and paper and prints [38]. Within this context, chelation agents are typically employed in one of three ways: the solubilization of otherwise insoluble metal compounds, sequestration to prevent metals from interfering with other cleaning processes, and the maintenance of specific types of ions in solution during cleaning [37].

Because most chelation is essentially an equilibrium type reaction, it is often largely dependent on the acidity/alkalinity of the solution and the concentration of the relevant reagents. The following discussion will therefore explore the chemical structure and solution kinetics of a few commonly used chelating agents.

*Composition*

Chelation gets its name from the Greek for claw (chele) due to the ability of these multi-dentate ligands to form coordinate bonds with a metal ion, essentially wrapping around the metal and forming a soluble complex [37]. The active sites are often acidic acetate groups organised along a flexible backbone which are sufficiently well-spaced to avoid any steric hindrance [39]. Furthermore, as not every chelate will complex every metal, there must also be electrochemical affinity between the compounds.

In other words, for a coordinate complex to form, the chelate must be both large enough to wrap around the ion and structured in a manner that allows it to bond from at least two directions [3]. The oxidation state of the ion must also be compatible with the chelate.

There are a large number of proprietary chelation agents, but specific discussion will be limited to the five compounds which appear most commonly in the conservation literature.

| | | |
|---|---|---|
| Citrates | Citrate anion | CAS 126-44-3 |
| | Citric acid | CAS 77-92-9 |
| | TAC triammonium citrate | CAS 3458-72-8 |
| DTPA | Diethylenetriaminepentaacetic acid | CAS 67-43-6 |
| EDTA | Ethylenediaminetetraacetic acid | CAS 60-00-4 |
| HEDTA | (Hydroxyethyl)ethylenediaminetriacetic acid | CAS 150-39-0 |
| NTA | Nitrilotriacetic acid | CAS 139-13-9 |

*Behaviours*

While this book has endeavoured to make things as clear and practical as possible, with chelation agents there's no shying away from the kinetics.

SITE ACTIVATION

Before a chelating agent will complex a metal ion, the 'teeth' must be prepared to bite. This is called site activation, and essentially describes how readily, and under what conditions the 'teeth' of the chelation agent will bond with the desired metal ion.

With acetate-based chelates often used in conservation, the formation of the metal-ligand complex depends on the coordinate bonding between the metal ion and several points on the chelate. The 'teeth' of the chelate must themselves be ionized, leaving a lone electron pair available for bonding before they can surround and complex the metal. The ionisation of these 'teeth' is called site activation.

Each activation site resides in its own electrochemical environment and is greatly affected by the behaviour of its neighbours. As a result, ionization is a step-wise process, with different dissociation requirements for different sites [3]. Site activation is most often a dissociation, with an acidic proton separating from its conjugate base. Therefore, site activation is nearly always described in terms of acid dissociation and the acid dissociation constant.

Generally speaking, the strength of a weak acid is quantified by comparing the relative concentration of an acidic proton ($H^+$) and its conjugate base (A-) at constant equilibrium (K). The greater the concentration of protons in relation to the base, the stronger the acid [3, 36]. This is known as the acid disassociation constant ($K_a$) and within the context of chelating agents it can be used to estimate how readily a given activation site will ionize.

As with pH, the equations used generate the acid dissociation constant yield very small numbers that are awkward to discuss; convention therefore dictates that this value be described in terms of the –log, or $pK_a$. For example, if an acid has a $K_a$ of −4, the $pK_a$ will be 4. The $pK_a$ can be therefore understood as inversely proportional to the strength of an acid, with small and negative values corresponding to the strongest acids [3, 36]. For chelation agents, the higher the $pK_a$, the stronger the affinity between proton and base and the less likely the site is to ionize.

For example, the first dissociation of EDTA happens readily, with a $pK_a$ of 1.99, while the fourth and final is 10.76 [40]. However, as this is a logarithmic scale, the fourth site is not 5.5 times less likely to dissociate; the affinity between acid and base is nine orders of magnitude, or nearly a billion times less, than the first.

COMPLEXING

Once the teeth are ready, with their lone electron pairs ready to bond with the metal ion, how quickly and how strongly this complexing actually occurs is due to another set of factors. This has to do with the physical size of the materials involved as well as the stability of the complex itself (as reflected in the formation constant).

As with the activation of the complexing sites, a similar stepwise process is involved in the formation of the metal complex itself. Once the active sites have been deprotonated and are available for bonding, coordination bonds will form between the chelate and the ion. However, just as dissociation energy was linked to site location, the stability of these bonds is contingent on local electrochemical effects and steric (spatial) considerations [3, 37]. The stability of complex coordination bonds is therefore also understood as an equilibrium process with its own constant (K). Instead of the interaction of acids and conjugate bases, the formation constant is an expression of the concentration of metal ions ($M^+$) and ligands (L-) in solution and is also typically cited in logarithmic form (log K).

Just as a high $pK_a$ corresponds to a strong bond between proton and base, a high log K value indicates a stable coordination bond [3].

For example, EDTA can bond most closely with metals it can fully envelop, such as nickel, while larger ions such as calcium stretch the preferred conformation and are therefore more weakly held [3]. Imagine nickel as a plum you can easily take a bit out of, while calcium would be like trying to bite into a bowling ball (silly analogy not to scale). This difference is borne out in formation constants for the two compounds and the log K of EDTA and calcium (16.11) is less than the log K for nickel (18.56) [3]. As kinetics favours the formation of the most stable products, if added to a solution of equal parts calcium and nickel, EDTA will complex a greater amount of the latter.

### PH AND BUFFERS

The acid dissociation and formation constants can help to predict the behaviour of chelating agents in solution, but the manipulation of pH can actually alter the behaviour of the system.

If the pH of a chelate-metal solution is too low, the acidic hydrogen ions bonded to the chelate's active sites will be stabilized in situ and will not readily dissociate, thus making the chelating agent unavailable for complexing [3]. Furthermore, if the proton (H+) does manage to break away, the negatively charged site will be quickly re-protonated by another available hydrogen ion before the metal ion can be complexed. If the pH goes down, acid-conjugate base affinity increases, and the $pK_a$ goes up.

Conversely if the system is too basic, the excess hydroxyl group will compete for bonding to the metal, inhibiting the formation of the complex or, even worse, precipitate with the metals directly, producing a network solid [41]. As the pH goes up, metal-ligand bonds are more difficult to form, and the value of log K goes down. Because both dissociation and formation constants are equilibrium systems, altering the concentration of protons is therefore able push the system towards one set of products or another [3].

As a result of this relationship, many (but not all!) chelating agents typically operate best in slightly alkaline conditions that are sufficient to encourage chelate ionization without competing for metal coordination. However, the exact value required depends on the agent being used and the ion being complexed [3].

The behaviours quantified in the aforementioned constants are described in terms of aqueous systems. However, the solubility of both chelation agent and complex are primarily determined by polarity, and as such chelation agents can operate in both aqueous and non-aqueous systems [3].The solubility of a complex is particularly important in chelation cleaning where the eventual removal of the metal ion is desired. The effectiveness of a chelating agent is therefore determined by the degree to which both metal and ligand are available for bonding and how easily the complex can then be removed.

It's important to remember that pH can shift during treatment and EDTA in particular can release H+ ions into solution, causing the acidity to increase. In many instances the metal replaces the proton in a complex, leaving the hydrogen free to associate with whatever anion in available (including those anions best left within in the object's surface).

*Characteristics of Select Sequestering Agents*

CITRATES

The term 'citrate' is used to describe several formulations of essentially the same acidic ligand (Figure 4.4). Citrates are typically provided as citric acid itself or one of three ammonium-citrate salts. Citric acid has three acidic sites, and in ammonium citrates, one, two, or three of these sites have been deprotonated and replaced with one, two, or three ammonium anions.

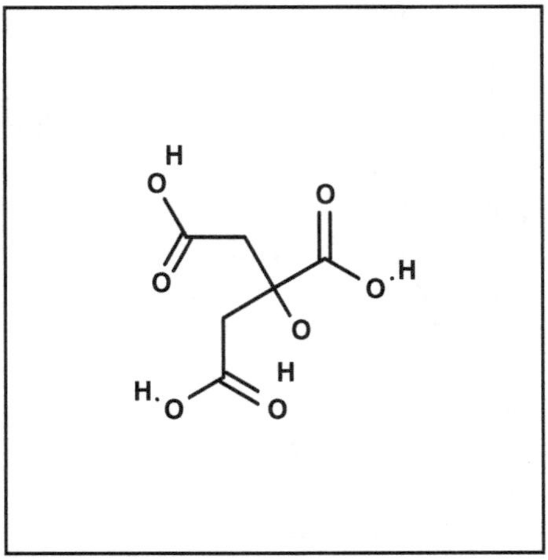


*Figure 4.4* Structure of Citric acid molecule.

Although mono- and di-ammonium citrates are in use, tri-ammonium citrate (TAC) is by far the most commonly employed. This is likely in no small part due to its ease of use, as all three active sites are fully ionised in solutions greater than pH 8. One would have to add additional ammonia to fully ionise mono or di-ammonium citrate at the same pH and the addition of different pH adjusting compounds can result in undesired compounds. As Gervais points out, the addition of sodium hydroxide to diammonium citrate will not yield triammonium citrate, but instead diammonium sodium citrate [42].

If using AC, one must be mindful that the ammonia will readily evaporate from solution, and this effect can drive up the pH of the system [43].

Research from paintings conservation has indicated that a single clearing of deionized water is sufficient to reduce AC levels to below GCMS detection levels, but only when applied in solutions less than 5%. More concentrated solutions require further rinsing, but how much has not yet been formally described [43].

### DTPA

DTPA, also known as pentetic acid, is not as widely used as citrates but has featured in a few investigations into paper foxing and paintings cleaning (Figure 4.5) [44–46].

DTPA is soluble in hot water and alkaline solution, but only slightly soluble in room temperature water. The maximum solubility for solids in 20°C water is roughly 5 g/L. It is insoluble in organic solvents such as alcohol and ether. DTPA is listed as being particularly effective in the removal of carbon-based materials, such as smoke residue or organic soiling [47].

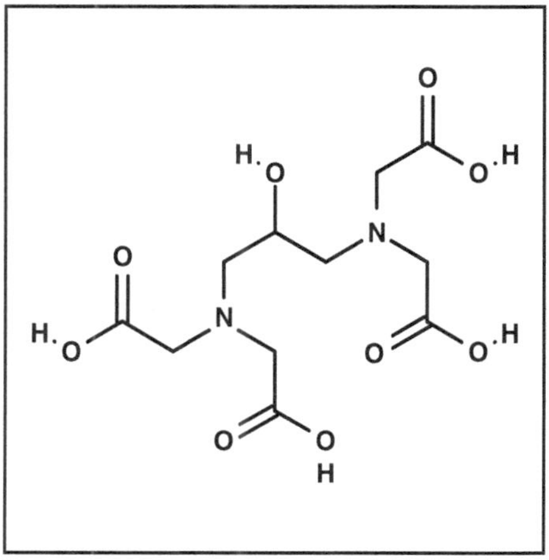

*Figure 4.5* Structure of DTPA molecule.

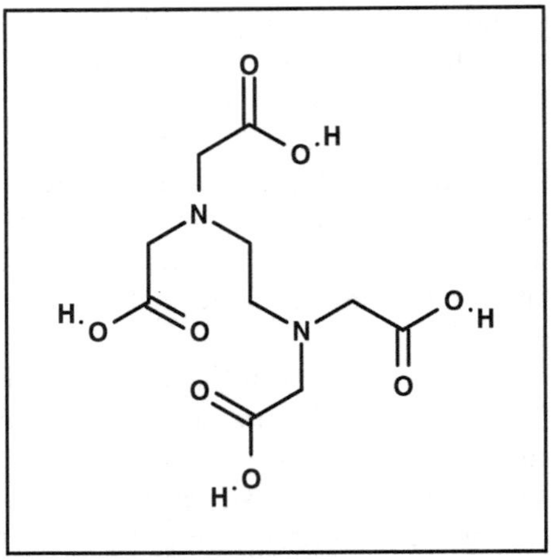


*Figure 4.6* Structure of EDTA molecule.

EDTA

EDTA, its salts, and related compounds are sometimes referred to as 'edetate' (particularly in the biomedical context) and it has been in common conservation use for years (Figure 4.6).

Disodium EDTA ($Na_2EDTA$) is acidic and when applied to a rust-stained marble surface, it will quickly solubilize calcite leaving the rust unaffected. This is despite the higher stability constant for iron complexes when compared to calcium. It is worth repeating this again, for although careful comparison of log K figures will show that $Na_2EDTA$ should preferentially complex iron over calcium, research has found the opposite to be true [33].

As the reactivity of chelation agents is a hugely influenced by solvent conditions, different EDTA salts will also have markedly different behaviours. While disodium EDTA will readily dissolve calcite, tetrasodium EDTA is found to preferentially attack gypsum and whewellite [33, 48].

A mechanism for this variation has not yet been proposed within the literature, but the variation found within type of EDTA underlines the importance of testing.

Finally, as EDTA will readily ionise its first few activation sites, this can result in an excess of free hydrogen cations in solution. This small detail can be of particular significance for those hoping to use sulphate-containing buffers as the combination of available hydrogen and $SO_4^{2-}$ anions essentially creates sulphuric acid. See 'Buffers and pH adjustment ' for more information.

HEDTA

There are no formal case studies featuring the use of HEDTA in the conservation literature, but it occasionally appears in blog posts, databases, and

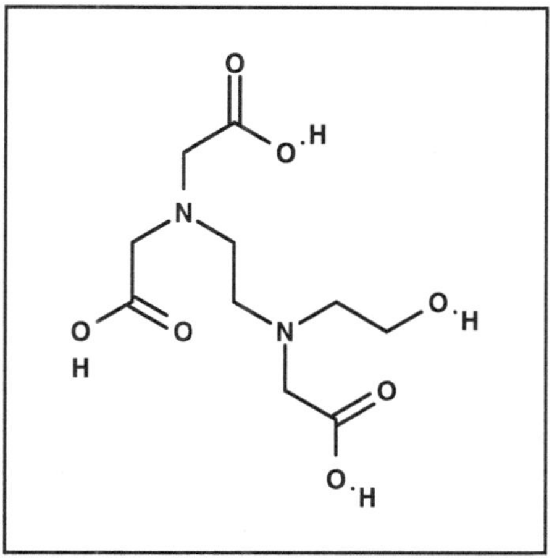


*Figure 4.7* Structure of HEDTA molecule.

informal discussion (Figure 4.7) [49]. The absence of literature could reflect either sparse use within conservation or use that is so routine as to be not worth mentioning.

NTA

A relatively weak chelating agent, NTA has seen use primarily for its ability to solubilise calcium deposits bound to residual soaps found in textiles or as surface coatings on stone (Figure 4.8). This weakness is a benefit as it has been found to be able to dissolve fatty acids bound with calcium salts while falling short of attacking many calcium-containing stone (of which marble is a prime example) [44, 50].

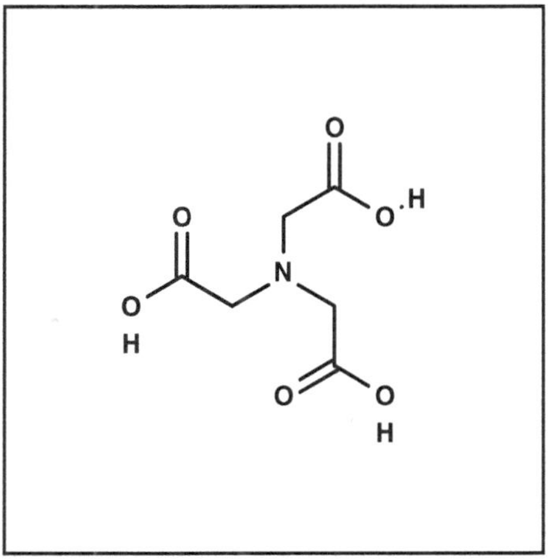


*Figure 4.8* Structure of NTA molecule.

*Key Ideas*

- Acid dissociation constant $pK_a$: for chelation agents, $pK_a$ is an expression of how tightly bound a base is to its proton. High value $pK_a$s correspond to tightly held protons that won't easily dissociate, while low $pK_a$ indicates that a site can be more easily deprotonated. In summary, the lower the $pK_a$ the easier it is to activate a chelation agent. It can be helpful to remember that the 'a' in $pK_a$ is for the acidic dissociation.

  For example: the $pK_a$ of the first ionisation site of EDTA is 1.99, and the $pK_{a1}$ of tri-ammonium citrate (TAC) is 2.79. In neutral conditions, EDTA will more readily shed its first acidic proton than TAC will.
- Dissociation constant, log K: Another equilibrium concept, here describing the stability of the ligand-metal complex. High numbers correspond to energetically preferred stable bonds, while lower numbers indicate less preferable bonding.

  For example: In a neutral solution, the log K of EDTA and $Cu^{2+}$ is 15.4 and the log K of EDTA and $Fe^{2+}$ is 10.9. In a solution pH 7 with equal concentrations of both metals, the EDTA is much more likely to complex copper than iron, because that complex is a more stable one.

## *Preparation: Sequestering Solutions*

### *Techniques and Tips*

As with the preparation of enzymes solutions, there are no special techniques to help with the creation of effective mixtures containing chelating agents. It is more a question of material selection and planning than complex preparation. The amount required must be calculated, the solution prepared, applied, and then any excess must be rinsed away.

#### STORAGE

While dry chelation agents can keep for long periods, once mixed, the solutions can quickly lose their reactivity [51]. Unless you're using them very frequently, most conservators would be best served making up solutions as they are required and correctly disposing of excess materials once treatment is complete.

#### PH

Careful manipulation of these solution pH allows conservators to selectively target particular metals, or even ions of a specific oxidation state, for chelation. This is especially useful in the cleaning of painted surfaces and porous stone where conservators must be careful to remove soiling ions while avoiding complexing metals in pigments, or attacking the stone itself [3, 51, 52].

The correct pH can be obtained through pH adjustment methods described in Chapter 3, and some solutions would benefit from the inclusion of buffers. Before starting, think about what exactly you are trying to do. Is the desired pH precise or in a narrow range? Will the process produce excess hydrogen or alkaline ions? If so, these are all likely to affect pH and the addition of a buffer should be considered.

#### APPLICATION METHODS

As with surfactants, there is a great deal of versatility in how one might choose to apply a given chelate. Because of their targeted reactivity, chelation agents can be added to immersion baths, swabbed across surfaces with a microfiber cloth, or applied with a gel or poultice. Please see Chapter 2 for further discussion of gel types.

#### REMOVAL/CLEARANCE

Although their reactivity can fade over time, chelating agents can continue to work as long as there is water present. To prevent any etching or undesired activity all remaining chelating agents must be cleared away from the surface following treatment. Research into painted surfaces has indicated that a single clearing of deionized water is sufficient to reduce TAC levels to below GCMS detection levels, but only when applied in solutions less than 5% [43]. The amount of rinsing required for other chelates has not yet been formally described, but two or three rinses in deionised water are often cited in literature [3].

### *Materials*

Volumetric measure: flask, graduated cylinder, etc
Analytical balance
Weighing boats
Glass, plastic, or magnetic stirrer (no metal)
Deionised water
pH adjustment solution (if required)
Pipette
Buffer (if required)

### *Instructions*

1. Calculate the amount of solute needed.
2. Fill the volumetric measure with roughly 1/3 the final volume of water,
3. Add the chelate to the volumetric container and stir well.
4. Fill the container to just short of the desired final volume.
5. Add buffer. See section on buffers for details.

6. Adjust pH. See sections on pH under 'Measurement' for methods and 'Calculations'.
7. Add water and pH adjustment solution until desired volume and pH is met.

### *Technical Information*

This section collates some usage guidance pulled from various sources in the academic literature. Technical values given below will be influenced by the methods and conditions of their measurement, so some variation within the literature is expected.

Although this is by no means comprehensive, it is intended to provide useful context for the early stages of material selection.

#### *Dissociation and Use*

*Table 4.3* pKa: Acid dissociation information for select chelation agents

| *Chelation agent* | *Solubility (sp) sparingly* | *pH range* | *pH (1 M)* | $pK_{a1}$ | $pK_{a2}$ | $pK_{a3}$ | $pK_{a4}$ | $pK_{a5}$ |
|---|---|---|---|---|---|---|---|---|
| Citric acid | Water alcohol (sp) | 3–6 | 3.13 | 4.76 | 6.40 | 4.76 | | |
| DTPA salt | Water | 2.1–2.5 | | 1.79 | 2.56 | 8.60 | 10.58 | 10.42 |
| EDTA salt | Water (sp) alcohol (sp) | 8 | 8 | 1.99 | 2.67 | 6.16 | 10.76 | |
| HEDTA salt | Water (sp) | 5–9 | 1.6 | 2.51 | 5.31 | 9.86 | | |
| NTA salt | Water (sp) Ethanol | | 1.65 | 2.94 | 10.33 | 9.73 | | |

Data collated from the National Institute for Standards and Technology (NIST) Chemistry WebBook SRD69 and Lange's Handbook of Chemistry [35, 40].

*Table 4.4* Selection of chelation agents and some of their uses in conservation

| *Chelate* | *Undesirable metal and substrate* | *Conc.* | *pH* | *Notes* | *Source* |
|---|---|---|---|---|---|
| AC | Grey deposit on painted canvas | 2.5% | 7 | Analysis of soiling indicated that it was a mixture of unknown organic material, Pb, Ca, C, S and P. | [53] |
| AC | Foxing on aged paper | 1–3% | 7–8 | Multiple float baths with rinsing between each. | [54] |
| AC | Copper staining on damask tablecloth | 0.02 M | 7 | pH adjustment and buffering with ammonium hydroxide | [55] |

(*Continued*)

*Table 4.4* (Continued)

| *Chelate* | *Undesirable metal and substrate* | *Conc.* | *pH* | *Notes* | *Source* |
|---|---|---|---|---|---|
| AC | No soiling removal, test for damage to marble surfaces | 2–5% | 9 | pH adjustment and buffering with ammonium hydroxide. Rinsed with calcium carbonate-saturated water. Lower concentrations and higher pH results in less likelihood of damage to marble | [42] |
| DTPA | Iron staining on paper | * | 9 | *Free acid of chelate added to 20% ammonium hydroxide until pH reached desired level.<br>Requires pre-soak in sodium dithionite ammonium solution to reduce iron to Fe (II) | [44] |
| DTPA | Surface soiling on painting | - | - | Within Wolbers' iso-solution D | [56] |
| EDTA | iron staining on paper | * | 9 | *Free acid of chelate added to 20% ammonium hydroxide until pH reached desired level.<br>Requires pre-soak in sodium dithionite ammonium solution to reduce iron to Fe (II) | [44] |
| EDTA | encrustations from archaeological ceramics | 10 | | Less effective than Na2EDTA | [57] |
| $Na^2EDTA$ | encrustations from archaeological ceramics | | | | [57] |
| NTA | Unknown oily surface coating on marble | 0.5% | 9 | Small amount of NaOH to dissolve and then pH adjusted back with HCl. Sodium borate buffer<br>Can solubilize calcium in fatty acids without destabilizing calcium carbonate | [50] |
| HBED* | Reduction in foxing on paper | 0.5% | 6 | With chitanase enzyme in agarose gel. Adjusted and buffered with citrates | [58] |
| HBED* | Reduction in foxing on linen | 0.5% | 6 | With proprietary enzyme 'Glucanex' in immersion baths and agarose gel. Adjusted and buffered with citrates | [59] |

* HBED (N,N'-Di(2-hydroxybenzyl) ethylenediamine-N,N'-diacetic acid – a strong iron chelator.

*Formation and Conditional Stability Constants*

The information provided is intended to help narrow down material selection, but these figures are not wholly predictive of all chemical behaviours. A good example of a predictive misfire is EDTA's aforementioned preference for calcite over rust as this affinity is not reflected in the stability constant of the complex. As always, careful testing is vital.

FORMATION CONSTANT

*Table 4.5* log K formation constants for some metal ions

| | *Citric acid* | *DTPA* | *EDTA* | *HEDTA* | *NTA* |
|---|---|---|---|---|---|
| CAS free acid | 77-92-9 | 67-43-6 | 60-00-4 | 150-39-0 | 139-13-9 |
| Alternate CAS | | | Disodium 139-33-3 | | Trisodium 5064-31-3 |
| $Al^{3+}$ | 7.0 | - | 16.11 | - | >10 |
| $Ba^{2+}$ | 2.98 | 8.8 | 7.78 | 5.54 | 5.88 |
| $Ca^{2+}$ | 4.68 | 10.9 | 11.0 | 8.43 | 7.60 |
| $Cd^{2+}$ | 3.98 | 19.0 | 16.4 | 13.0 | 9.80 |
| $Co^{2+}$ | 4.8 | 19.0 | 16.31 | 14.4 | 10.38 |
| $Cu^{2+}$ | 4.35 | 21.1 | 18.7 | 17.40 | 13.10 |
| $Fe^{2+}$ | 3.08 | 16.0 | 14.33 | 11.6 | 8.84 |
| $Fe^{3+}$ | 12.5 | 27.5 | 24.23 | 19.8 | 15.87 |
| $Hg^{2+}$ | - | 27.0 | 21.8 | 20.1 | 12.7 |
| $Mg^{2+}$ | 3.29 | 9.3 | 8.64 | 5.78 | 5.36 |
| $Mn^{2+}$ | 3.67 | 15.5 | 13.8 | 10.7 | 8.60 |
| $Ni^{2+}$ | 5.11 | 20.2 | 18.56 | 17.0 | 11.26 |
| $Pb^{2+}$ | 6.5 | 18.8 | 18.3 | 15.5 | 11.8 |
| $Sr^{2+}$ | 2.8 | | 8.80 | 6.92 | 6.73 |
| $Zn^{2+}$ | 4.71 | 18.4 | 16.14 | 14.5 | 10.45 |

Higher log K values generally refer to more stable complexes. Data for citric acid, EDTA, HEDTA, and NTA collated from Table 8.13 in *Lange's Handbook of Chemistry* [35]. Data for DTPA from Rivers and Umney (2013) [5].

CONDITIONAL STABILITY CONSTANTS

Where available, the conditional stability constant is given for several chelates. These figures are not a true constant as the value can change based on experimental conditions. However, it more closely (albeit still not perfectly) predicts the order in which a given chelate will react with different metal ions at a specific pH. Unfortunately, the data set below does not include any citrates or NTA. All data from Dow's *Chelation Chemistry'*[36] (Tables 4.6–4.8).

*Table 4.6* EDTA – Conditional stability constants

| | *pH* | | | | | | | | | | |
|---|---|---|---|---|---|---|---|---|---|---|---|
| | *2* | *3* | *4* | *5* | *6* | *7* | *8* | *9* | *10* | *11* | *12* |
| $Al^{3+}$ | 2.8 | 5.2 | 7.2 | 8.1 | 7.2 | 5.9 | 3.9 | 3.4 | 1.0 | | |
| $Ba^{2+}$ | | | | 1.3 | 3.0 | 4.4 | 5.5 | 6.4 | 7.3 | 7.7 | 7.8 |
| $Ca^{2+}$ | | | 2.2 | 4.1 | 5.9 | 7.3 | 8.4 | 9.3 | 10.2 | 10.6 | 10.7 |
| $Cd^{2+}$ | 3.8 | 6.0 | 7.9 | 9.9 | 11.7 | 13.1 | 14.2 | 15.0 | 15.5 | 14.4 | 12.0 |
| $Co^{2+}$ | 3.7 | 5.4 | 7.8 | 9.7 | 11.5 | 12.9 | 13.9 | 14.5 | 14.7 | 14.0 | 12.1 |
| $Cu^{2+}$ | 6.1 | 8.3 | 10.2 | 12.2 | 14.0 | 15.4 | 16.3 | 16.6 | 16.6 | 16.1 | 15.7 |
| $Fe^{2+}$ | 1.5 | 3.7 | 5.7 | 7.7 | 6.5 | 10.9 | 12.0 | 12.8 | 13.2 | 12.7 | 11.8 |
| $Fe^{3+}$ | 11.5 | 13.9 | 14.7 | 14.8 | 14.6 | 14.1 | (13.7) | (13.6) | (14.0) | (14.3) | (14.4) |
| $Hg^{2+}$ | 9.2 | 11.1 | 11.3 | 11.3 | 11.1 | 10.5 | 9.6 | 8.8 | 8.4 | 7.7 | 6.8 |
| $Mg^{2+}$ | | | | 2.4 | 3.9 | 5.3 | 6.4 | 7.3 | 8.2 | 8.5 | 8.2 |
| $Mn^{2+}$ | 1.4 | 3.5 | 5.5 | 7.4 | 8.2 | 10.6 | 11.7 | 12.6 | 13.4 | 13.4 | 12.6 |
| $Ni^{2+}$ | 6.1 | 8.2 | 10.1 | 12.0 | 13.8 | 15.2 | 16.3 | 17.1 | 17.4 | 16.9 | 16.0 |
| $Pb^{2+}$ | 5.2 | 7.4 | 9.4 | 11.4 | 13.2 | 14.5 | 15.2 | 15.2 | 14.8 | 13.9 | 10.6 |
| $Sr^{2+}$ | | | | 2.0 | 3.8 | 5.2 | 6.3 | 7.2 | 5.1 | 8.5 | 8.6 |
| $Zn^{2+}$ | 3.8 | 6.0 | 7.9 | 9.9 | 11.7 | 13.1 | 14.2 | 14.9 | 13.6 | 11.0 | 8.0 |

*Table 4.7* DTPA – Conditional stability constants

| | *pH* | | | | | | | | | | |
|---|---|---|---|---|---|---|---|---|---|---|---|
| | *2* | *3* | *4* | *5* | *6* | *7* | *8* | *9* | *10* | *11* | *12* |
| $Al^{3+}$ | | 2.5 | 4.9 | 6.5 | 6.2 | 5.5 | 3.0 | 3.0 | | | |
| $Ba^{2+}$ | | | | | 1.3 | 3.5 | 5.5 | 7.1 | 8.1 | 8.7 | 8.8 |
| $Ca^{2+}$ | | | 1.5 | 3.0 | 4.1 | 5.7 | 7.6 | 9.2 | 10.2 | 10.8 | 10.9 |
| $Cd^{2+}$ | 2.1 | 5.1 | 7.5 | 9.7 | 11.7 | 13.7 | 15.7 | 17.2 | 17.8 | 16.9 | 14.5 |
| $Co^{2+}$ | 3.0 | 5.9 | 8.1 | 9.9 | 11.7 | 13.7 | 15.6 | 16.9 | 17.2 | 16.7 | 14.8 |
| $Cu^{2+}$ | 5.3 | 8.2 | 10.3 | 12.1 | 13.8 | 15.8 | 17.6 | 18.6 | 18.7 | 18.3 | 17.4 |
| $Fe^{2+}$ | 0.6 | 3.5 | 5.6 | 7. | 8.8 | 10.7 | 12.7 | 14.5 | 15.7 | 16.4 | 16.5 |
| $Fe^{3+}$ | 10.1 | 12.7 | 14.0 | 14.5 | 14.5 | 14.5 | (14.5) | (14.2) | (13.5) | (12.8) | (11.9) |
| $Hg^{2+}$ | 9.8 | 12.3 | 13.4 | 13.8 | 13.8 | 1.38 | 13.8 | 13.4 | 12.5 | 11.0 | 9.1 |
| $Mg^{2+}$ | | | 0.4 | 1.9 | 3.0 | 4.3 | 6.0 | 7.6 | 8.6 | 9.1 | 8.8 |
| $Mn^{2+}$ | | 2.1 | 4.3 | 6.3 | 8.2 | 10.2 | 12.2 | 13.8 | 14.7 | 14.9 | 14.1 |
| $Ni^{2+}$ | 5.0 | 7.9 | 10.0 | 11.6 | 13.0 | 14.9 | 16.9 | 18.4 | 18.8 | 18.5 | 17.6 |
| $Pb^{2+}$ | 2.5 | 5.4 | 7.6 | 9.6 | 11.5 | 13.4 | 15.0 | 15.7 | 15.4 | 14.0 | 11.4 |
| $Sr^{2+}$ | | | | 0.9 | 2.5 | 4.4 | 6.4 | 8.0 | 9.0 | 9.6 | 9.7 |
| $Zn^{2+}$ | 3.2 | 6.1 | 8.2 | 6.7 | 11.2 | 13.1 | 15.1 | 16.5 | 15.3 | 12.9 | 9.9 |

*Table 4.8* HEDTA – Conditional stability constants

| | *pH* | | | | | | | | | | |
|---|---|---|---|---|---|---|---|---|---|---|---|
| | *2* | *3* | *4* | *5* | *6* | *7* | *8* | *9* | *10* | *11* | *12* |
| $Al^{3+}$ | | | | | | | | | | | |
| $Ba^{2+}$ | | | | 0.9 | 2.3 | 3.4 | 4.4 | 5.3 | 6.0 | 6.2 | 6.2 |
| $Ca^{2+}$ | | | 1.2 | 3.1 | 4.5 | 5.6 | 6.6 | 7.5 | 8.2 | 8.4 | 8.4 |
| $Cd^{2+}$ | 1.0 | 3.6 | 5.8 | 7.7 | 9.1 | 10.2 | 11.2 | 12.0 | 12.3 | 11.0 | 8.5 |
| $Co^{2+}$ | 2.4 | 5.0 | 7.2 | 9.1 | 10.5 | 11.6 | 12.5 | 13.1 | 13.1 | 12.2 | 10.2 |
| $Cu^{2+}$ | 5.4 | 8.0 | 10.2 | 12.1 | 13.5 | 14.6 | 15.4 | 15.7 | 15.5 | 14.7 | (13.7) |
| $Fe^{2+}$ | 0.2 | 2.8 | 5.0 | 6.9 | 8.3 | 9.4 | 10.4 | 11.5 | 12.4 | 12.7 | 12.7 |
| $Fe^{3+}$ | 7.8 | 10.1 | 11.2 | 11.9 | 12.3 | (12.4) | (12.4) | (12.3) | (12.0) | (11.2) | (10.2) |
| $Hg^{2+}$ | 8.1 | 10.2 | 11.0 | 10.9 | 10.3 | 9.4 | 8.4 | 7.3 | 6.0 | 4.2 | 2.2 |
| $Mg^{2+}$ | | | | 0.5 | 1.9 | 3.0 | 4.0 | 4.9 | 5.6 | 5.7 | 5.3 |
| $Mn^{2+}$ | | 1.3 | 3.5 | 5.4 | 6.8 | 7.9 | 8.9 | 9.8 | 10.4 | 10.2 | 9.3 |
| $Ni^{2+}$ | 5.0 | 7.6 | 9.8 | 11.7 | 13.1 | 14.2 | 15.2 | 16.0 | 16.1 | 15.4 | 14.4 |
| $Pb^{2+}$ | 3.5 | 6.1 | 8.3 | 10.2 | 11.6 | 12.6 | 13.2 | 13.2 | 12.6 | 10.8 | (8.1) |
| $Sr^{2+}$ | | | | 1.5 | 2.9 | 4.0 | 5.0 | 5.9 | 6.6 | 6.8 | 6.8 |
| $Zn^{2+}$ | 2.5 | 5.1 | 7.3 | 9.2 | 10.6 | 11.7 | 12.7 | 13.4 | 11.9 | 9.1 | (6.0) |

## ENZYMES

Consultants: Carola Hilby and Joanne Hackett

### *Material Information*

Precision Required: High

#### *Caution*

As with all materials in the 'cleaning' section, enzymes and their use come with a few strong words of warning. While they are incredibly selective in the chemical bonds they target, they are not selective of its source. An amylase used to break down unwanted starch paste will simultaneously quite happily attack any starch size within the paper. Enzymes require the presence of water and can be reactivated long after the initial application if not carefully washed away. They can also be prohibitively expensive, fussy to work with, or require conditions which are unsuitable for heritage treatments.

#### *Overview*

Enzymes are essentially biological catalysts. Their complex three-dimensional shapes allow them to interact with specific sites on specific molecules in a manner that encourages specific reactions [60]. From a conservation perspective, this precision can be exploited to target particular chemical bonds within an unwanted material, thereby permitting highly controlled removal.

As with chelation, this section will serve to outline some of the main ways in which enzymes have been used in conservation, providing a starting point to investigations for use. They can be prohibitively expensive, and where possible that information has been included. However, as with everything in this book, specific details are likely to change swiftly.

Finally, it's worth mentioning that there has been significant work in the development of live bacteria facilitated cleaning [61–66]. At the moment, the preparations are a bit too complex for the aims of this text, but this marks a promising area of future research and development. Interested parties are encouraged to see Ranalli and Zanardini for an excellent literature survey on the topic [67].

*Composition*

The chemical composition of enzymes is broadly known, consisting of various combinations of amino acids into long polypeptide chains with the odd smattering of side chains. These chains are then folded and twisted onto themselves to form an extraordinarily complex spatial arrangement. It is this shape which provides much of their catalytic specificity and which has been the subject of significant scientific interest for decades [60]. As with many proteins, the details of their structure is enormously, staggeringly, complex, difficult to model, and from a conservation point of view, of little practical use. However, this complexity results in a chemical precision unmatched within other conservation materials.

The most often employed analogy is that of a lock and key, where the shape of the folded enzymes perfectly matches the substrate that is acted upon. It involves bringing two materials into physical proximity in a way that lowers the activation energy for the creation of the reaction product [60]. While the reality is, as ever, more complicated that this implies, it does communicate the selectivity of these pairings.

As the relationship between enzyme and substrate is dependent on complex spatial coordination of a folded peptide chain, anything that can affect that shape, such as temperature or pH, will affect whether the reaction will occur.

*Classifications*

NOMENCLATURE

Distinct from other chemical naming conventions, enzyme nomenclature is organized by behaviour rather than composition. Colloquial names for enzymes are typically based on the material or bond type that they coordinate with, and the suffix '–ase'. For example, cellulase catalyses the cleaving of cellulose and protease will facilitate the breakdown of proteins [60, 68, 69]. As understanding of these molecules has improved, new, more systematic identification approaches have been brought into use.

The Enzyme Commission or EC number is perhaps the simplest and most common. The EC number aligns with previous naming systems, classifying in terms of the reaction a given enzyme catalyses offers both more flexibility and precision. Each enzyme is mapped to a four-component number which exactly describes the type, location, and mechanism of the coordinated reaction. For example α-amylase (EC 3.2.1.1) and β-amylase (EC 3.2.1.2) both cleave the α-(1-4) glycosidic bond found in starch, but α-amylase attacks each such bond in turn, one after another, while β-amylase leapfrogs, cleaving at every-other instance [70, 71]. As a result, the product of α-amylase reaction is single rings of glucose, while the product of β-amylase is the disaccharide maltose. This difference is reflected in their EC numbers with identical reaction types, subclasses, and sub-subclasses, but slightly different activity resulting in near, but not quite identical, EC numbers [70]. This is actually an incredibly handy identification method as it reflects that key enzymatic specificity, but also allows the conservator some advantage in selecting an enzyme from among the dizzying array of available options.

Finally, a vast majority of the enzymes used in conservation are of the hydrolase class, the third of seven major enzyme types. This can be seen with the leading '3' in each EC number given below. Further exploration into other bioactive materials and other enzyme classes will no doubt prove fruitful avenues for future research [61, 72–75].

### SOURCE

The simplicity and clarity of the EC system might encourage someone to assume that given their shared EC number, all α-amylase are essentially the same. Nothing biological is that simple, and enzymes evolved and adapted within organisms to address issues specific to those organisms. The source of an enzyme can therefore affect many characteristics and behaviours, such as pH and temperature optima as well as relative activity levels [60]. Within each enzyme type there are numerous sources with some are derived from animal pancreas, saliva, or other digestive systems, while others are extracted from bacteria, fungi, or microorganisms. An α-amylase derived from *Bacillus substilis* will likely have slightly different activity, pH, and temperature optima from α-amylase from *Bacillus licheniformis*, but both are EC 3.2.1.1 and both will cleave the exact same 1-4 glycosidic bond discussed above.

Enzyme source therefore affects the optimum solution conditions for an enzymatic cleaning system, which, while difficult to sometimes unpick, allows for greater flexibility in treatment design. If one protease requires higher pH than is safe for a specific object, then the conservator can always investigate alternate protease of the same EC but from a different source. Tailoring enzyme selection to the exact needs of the treatment.

*Quantity*

ACTIVITY

As enzymes are valued for their ability to catalyse reactions, and catalytic activity is linearly dependent on the concentration of enzymes present, they are often described in terms of *activity*. Activity is confusingly described in units of 'unit', and one unit of activity is defined as the amount required to catalyse 1 µmol of substrate per minute under optimised standard conditions. As with any such system, there is some wiggling and manoeuvring in translating what is 'optimal' for a given assay method to what is practical and suitable, but activity remains a useful metric for comparing how quickly and effectively enzymes work [60].

While greater specificity and precision is attainable with labs which are equipped to conduct their own assays, this is simply not practical for most. However, so long as *high-quality* enzymes, bought from reputable sources, and good lab practice is maintained during storage and decant, the given activity units should reflect practical use [69].

While activity can dimmish over time, how much and how long has not been formally determined. Erickson suggests that conservators check whether enzymes are still working by simply preparing samples and informally testing the results. For starch-eating amylase, this can involve adhering some strips of paper together with wheat starch paste, warming the samples in a humid oven (90–98 °C) for 3–7 days to provoke cross-linking, and then testing the enzyme solution against a control [69].

CONCENTRATION

The higher the activity of enzyme, the more units per volume, the lower the amount of enzyme required to produce an effective solution. Theoretically, then, under the banner of minimal intervention, conservators should aim to make very dilute enzyme solutions of highly active enzymes. Lower concentrations also require less rinsing to remove or clear the enzymes when the treatment aims have been met. Concentration is therefore a question of activity in solution, measured as units/mL for immersion bath systems. In the specific instance of α-amylase, 1–5 units of activity per mL is ideal, but this can inch up closer to 50–250 u/mL without creating excess clearance issues [69]. It's worth noting, however, that there are a handful of proprietary activity units, such as kilo novo units (KNU) used by Novozyme, which cannot be easily converted. The activity for these must be drawn directly from manufacturer information and evaluated independently [76].

However, a quick aside that these very small volumes can exceed the precision limitations of some bench-top weighing devices. Conservators working without access to high precision analytical balances should consider making larger stock solutions and then diluting down when required. Manufacturers will provide storage information for each of their enzymes, but needless to say, they become less effective over time. Conservators using stock solutions can therefore expect some amount of waste, with the exact percentage dictated by the nature and frequency of enzymatic treatments being conducted.

PURITY

This can be a potentially muddy topic given the ambiguity over how to define 'purity' and the complexity of determining how pure is pure enough. It is, however, worth mentioning two points about the presence of unknown materials in the enzyme and the treated substrate. It's reported that isolating the desired amylase from the very much undesired cellulase can be technically challenging. When working with paper or other cellulosic materials, it is therefore important to work with highly refined amylase with minimal side activity. Conversely, substrates used for assays are highly controlled, and therefore don't reflect the complexity of a historic surface. One can therefore expect that the activity during treatment will be less than formally described because dust, metal ions, and other compounds on the object surface can inhibit enzyme activity [76].

***Preparation: Enzymes***

*Techniques and Tips*

Beyond meeting the specific pH and temperature conditions for a given enzyme, there is no special technique for enzyme solution preparation. The amount required must be calculated, dissolved, applied, and then any excess must be rinsed away.

That said, the successful preparation of enzymes requires good adherence to laboratory best practices. Please see the section on 'Mixing' for further information about material interactions.

SOURCING QUALITY MATERIALS

Given the variation and expense in purchasing enzymes, there is no small temptation to save money by buying less-effective, lower-quality materials. However, these can contain any number of unlisted bulking agents, metal ions, and other contaminating materials which can significantly affect the activity of the enzymes, but can also bond with the object's surface, leaving difficult-to-remove residues. Look for high activity, 'cold-active' enzymes which show good activity at object appropriate temperatures which are supplied as lyophilized (freeze-dried) powders [77]. Very high-quality enzymes are also supplied in ampoules, but these can be cost-prohibitive for many applications.

It's also worth mentioning that some researchers have had success obtaining 'small' samples directly from enzyme production companies. The relativity of what makes for a 'small' sample to a company accustomed to selling literal metric tonnes of materials means that some very generous samples (250–500 mL) can be obtained. This is a particularly promising avenue for students and those who can communicate a strong argument for the altruism of heritage conservation.

### ENZYME COMBINATIONS

Mixtures of different enzymes can be used when a combination of activity is desired. These can be both homemade bespoke mixtures as well and proprietary combinations, such as Glucanex, which integrates glucose, cellulase, protease, and chitinase [59]. However, if using a homemade combination, ensure that the enzymes are compatible. For example, if using protease and amylase to break down a mixture of animal glue and starch paste, remember that the protease will also attack the amylase, because it too is a protein. This will result in a reduction of activity for both enzyme types. These issues, however, are easily resolved by applying the materials in series rather than simultaneously [76].

### CALCULATING QUANTITY

Enzymes are typically supplied as either solids or pre-mixed liquids. If solids are used, ensure that your calculations reflect the activity given in the accompanying literature. Activity can be recorded as units per mg *solids*, units per mg *enzyme*, or another difficult-to-translate proprietary unit. If it's the lattermost, the only advice is to rely on the information provided by the manufacturer. For the former, units per solid, concentration for enzyme per mass of solid will need to be taken into account before further calculations are done [69]. Discussed with examples in Chapter 1.

$$m_s = \frac{C_V \times V}{A_s}$$

$m_s$ = the mass of solid enzyme-containing materials desired
$C_V$ = the concentration (units/mL) desired activity
$V$ = volume of the solution
$A_s$ = the units of activity per mg of solid

### MIXING

The choice of mixing vessel is very important with enzymes, as they are known to bond to the surface of glass, quartz, and polystyrene. This undesired bonding reduces the number of enzymes which are therefore available to work in the conservation treatment. Polypropylene (PP) and polyethylene (PE) are better options for mixing vessels, pipettes, and anything else coming into direct contact with the enzymatic material [77].

Particularly in the preparation of enzyme gels or thickened solutions, avoid over-mixing and the inclusion of air bubbles. Over-exposure to air can cause the enzymes to denature, diminishing their efficacy [78].

Water should be brought to desired temperature and pH adjusted before adding the enzyme powder or solution. Please note that although the activity may be less than 'ideal', many enzymes work well at neutral pH conditions. If the substrate is sensitive to pH, it is no doubt preferable to work at the acidity/alkalinity that is better for the object than the enzyme. Further specifics can be found in the technical data at the end of this section.

#### APPLICATION METHODS

The application method varies with the object type and degree of wet cleaning the object can withstand. Paper and textiles are sometimes immersion washed in enzyme solutions, but rigid gels have also proven an effective delivery method [79–81]. Paintings conservation often uses both thixotropic, soft, spreadable gels, as well as rigid gels [3, 82]. These delivery methods are covered in greater detail in Chapter 3. However, it is worth noting here that recent research indicates that enzyme solutions brushed onto a rigid gel surface are more effective than those incorporated within the body of the network [83].

#### REMOVAL/CLEARANCE

When enzymes were first being explored for conservation treatments, there was significant discussion about how to ensure their removal. While some efforts were made to denature the enzymes in situ, subsequent research has shown that this approach is more likely to cause deposition. Studies in the use of enzymes in paper and textile conservation have shown that two rinses in water are sufficient to remove enzymes and their products. While both of these studies employ immersion rinses, swabs of deionized water are also shown to be an effective method of removal [83–85].

### *Materials*

As discussed in 'Mixing', polyethylene and polypropylene materials are preferable for any items which may come into direct contact with dried enzymes or enzymatic solution.

Clean and dry (ideally sterile) lidded container
Thermometer or thermocouple
Volumetric measure: flask, graduated cylinder, measuring cups, etc.
Warming mat, hot plate, bain-marie, sous vide bath, or other warming device.
If supplied as solid: analytical balance and weighing boats
If supplied as liquid: micropipette dispenser and sterile tips for use with enzymes.
Stirring rod or mixing spatula

*Instructions: Simple Solution*

If using in a gelled system also see notes in the recipes for 'Spreadable Gels' and 'Non-spreadable Gels' in Chapter 3.

1. Calculate the amount of enzymes required using the activity information given by the supplier and the desired final volume of solution,
2. In a clean graduated cylinder, measure the total desired amount of water and pour into a clean, lidded container.
3. Set the container on or in the chosen warming method and bring to the correct temperature for enzyme activity.
4. Add buffering agents and adjust pH as required.
5. Add enzymes to solution and mix by gently stirring or swirling. Avoid vigorous actions which will incorporate air, as this can cause enzymes to denature.

***Technical Information***

*Table 4.9* Commonly used enzyme classes, their use and mechanism

| *E.C.* | *Name* | *Type* | *Breaks down* | *Mechanism* |
|---|---|---|---|---|
| 3.2.1.1 | α-amylase | carbohydrase | Starch | Cleave each α-(1-4) glycosidic bond, breaking down to glucose. |
| 3.2.1.2 | β-amylase | carbohydrase | Starch | Cleave alternating α-(1-4) bonds, breaking down to maltose. Less commonly used because maltose can be difficult to remove |
| 3.1.1.3 | triacylglycerol lipase | lipase | Paraloid B72; oily soiling | Cleaves carbolic ester in triglycerides and then diglycerides. |
| 3.4.23.1 | pepsin | protease | Animal glue | Cleaves hydrophobic, preferably aromatic, residues in P1 and P1' positions of amino acids |
| 3.4.21.4 | trypsin | protease | Animal glue | Cleaves peptides of lysine and arginine amino acids |
| 3.2.1.14 | chitinase | glycosidase | Chitin* in fungal cell walls | randomly cleaves glycosidic linkages in chitin and chitodextrins. |

Data collated from the BRaunschweig ENzyme DAtabase (BRENDA) [86].

* Chitin is an extremely abundant polysaccharide and forms a major structural component of insect exoskeletons and the cell walls of some fungi.

*Table 4.10* A selection of enzymes and their use in conservation

| *Type* | *Source* | *Use* | *Application technique* | *Activity* | *T* | *pH* | *Buffer* | *Quantity* | *Citation* |
|---|---|---|---|---|---|---|---|---|---|
| α-amylase | Bacillus sp. | Starch on textile | Gel surface, direct pipette | ≥1500 u/mg | | 7.3 | HEPES | 3 mg/mL | [83] |
| α-amylase | - | Starch on paper | Gellan gum and calcium acetate hydrogel | 30500 u/ml | - | 9 | PIPES | | [87] |
| α-amylase | Bacillus sp. | Starch on textile | In gel | ≥1500 u/mg | 40 | 6.9 | Sodium phosphate | 0.1% w/v | [88] |
| α-amylase | Bacillus sp. | Starch on textile | Immersion with Au nanoparticles | ≥1500 u/mg | | 7.3 | HEPES | 5 mg/mL | [75] |
| α-amylase | Aspergillus oryzae | Starch on paper | Immersion with Au nanoparticles | 30 u/mg | 25 | 6 | | | [73] |
| α-amylase | Aspergillus oryzae | Starch on textile | Immersion | | 40 | 6.9 | Sodium phosphate | 25 u/mL | [89] |
| α-amylase Amplify 12L | Bacillus lichenoformis | Wheat starch on paper. Test of iMat heating method only | 1.5 wt % agarose gel. | 12 SNU-A/g | 45–55 | 6 | - | 0.5–1% | [56] |
| α-amylase BAN 480L | Bacillus amyloliquefaciens | Wheat starch on paper. Test of iMat heating method only | 1.5 wt % agarose gel. | | 45 | | - | 0.025% | [56] |

| | | | | | | | | | |
|---|---|---|---|---|---|---|---|---|---|
| α-amylase | Bacillus amyloliquefaciens | Wheat starch on paper | 2% w/v agarose gel | 480 KNU | 20 | 7 | None | 1.5% (w/w) | [90] |
| α-amylase BAN 840L | Bacillus amyloliquefaciens | Wheat starch and glue on paper | Loaded into thick filter paper and built into a 'sandwich' with barrier layers and aluminium foil. | 480 KNU | 30–40 | 7 | None | 1.5% (w/w) | [91] |
| Chitinase | - | Foxing on paper | Agarose gel | 100 u/L | | 6 | Sodium citrate | 1% (w/v) | [58] |
| Protease (Trypsin) | Porcine pancreas | Animal glue on paper | Immersion | 15,200 u/mg | RT | 7.5 | Calcium hydroxide | 200 u/mL | [79] |
| Protease (Pepsin) | | Animal glue on paper. Note pH far too low for most paper treatments | Immersion | 4500 u/mg | RT | 3.4 | 0.5% potassium phosphate 0.5% citric acid | 200 u/mL | [79] |
| Lipase | Candida cylindracea | Paraloid B72 on oil on canvas | Thickened with methylcellulose, applied and agitated with brush | | 30 | 8 | Tris base; tris base hydrochloride | 0.005 g/mL | [92] |
| Lipase type VII | - | Oil on silk | Methyl cellulose spreadable gel | - | - | 8.5 | TRIS | 0.5% | [78] |
| Protease | | Animal products in paint on stone | Poultice | 200 u/mg | 42 | 9.5 | | 1.92 mg/mL | [93] |

(*Continued*)

*Table 4.10* (Continued)

| *Type* | *Source* | *Use* | *Application technique* | *Activity* | *T* | *pH* | *Buffer* | *Quantity* | *Citation* |
|---|---|---|---|---|---|---|---|---|---|
| Protease | Bacillus Licheniformis, Type VIII | Bat blood on painted surface and inscriptions | Applied with soft brush | 15 u/mg | 37 | 7.5 | | | [94] |
| Protease Alcalase 2.5L DX | Bacillus licheniformis | Rabbit skin glue on paper. Test of iMat heating method only | 1.5 wt % agarose gel | 2.4 AU/ml | 45 | 7 | - | 0.1% v/v | [56] |
| protease | Aspergillus Oryzae P-4755 Type II | Animal glue on silk and linen | Immersion bath | 20 u/mL | 37 | 7.5 | sodium acetate | | [95] |
| Proteinase | Tritirachium album | Glue on paper | Gellan gum and calcium acetate hydrogel | ≥30 U/mg | - | 8 | PIPES | | [96] |
| Glucanex (β-glucanase, cellulase, protease, and chitinase) | proprietary lysing enzyme | Foxing on linen | Series of immersion baths | - | 25–27 | 6 | Sodium citrate | 5 g/L | [59] |

## (MICRO)EMULSIONS

Consultants: Steven Abbott and Megan Creamer

***Material Information***

Precision Required: Medium

*Introduction*

The use of emulsions in conservation is particularly well developed within the cleaning of paintings, and several well-established scholars – such as Baglioni, Wolbers, and Stavroudis (to name a few) – have contributed a huge amount of literature on the topic. However, despite this original specificity, emulsions have broad applications in surface cleaning for many material types and provide a method which requires much smaller volumes of potentially harmful solvents. Both emulsions and microemulsions are discussed in parallel as their preparation largely overlaps.

*Overview*

Emulsions are a dispersed solution of at least two immiscible liquids, such as oil and water, by means of physical agitation (standard emulsion) or an intermediary compound, typically a surfactant. Microemulsions are *thermodynamically* stable and will form more or less spontaneously, without the high energy, high sheer mixing required by the *kinetically* stable emulsions and nanoemulsions [97]. In practice this means that emulsions will eventually separate into their constituent layers, while microemulsions are a single stabilised system that will not spontaneously re-divide.

The formation of both emulsions and microemulsions can be enabled by the dual character of surfactants. The ability of surfactants to solubilize unlike materials is exploited to create mixtures where non-polar, oily substances can be dispersed within polar, often aqueous liquids, and vice versa.

There are three common classifications of dispersions: Type I oil-in-water (O/W), Type II water-in-oil (W/O), Type III microemulsion. Within the first two types, the larger, latter half of each pair is called the continuous phase, and the smaller is the dispersed phase [6]. Broadly speaking, in order for most emulsions to form, the surfactant needs to be more soluble in the continuous phase than the dispersed. Although the difference between emulsions and microemulsions can perhaps seem a bit academic, it can typically be visually determined. Microemulsions are macroscopically homogeneous and often transparent, while emulsions are typically cloudy or turbid [97].

Another method of emulsification is found in the Pickering emulsions, which utilize solid particles as stabilisers rather than surfactants. These particles accumulate at the interface between the two immiscible liquids, forming pockets which can then be 'activated' or ruptured by gentle agitation such as the action of spreading with a soft brush [98–100]. The preparation of Pickering systems, although technically emulsions, share more in common with the thixotropic gels and Laponite systems which are discussed in Chapter 3 [101]. As such, the rest of this section will focus on the preparation of surfactant-mediated emulsions.

### *Behaviours and Characteristics*

#### HYDROPHILIC-LIPOPHILIC BALANCE (HLB) AND SHORTCOMINGS

Broadly speaking, water-soluble surfactants will form oil-in-water microemulsions (O/W) and oil-soluble ones will bring aqueous solutions into non-polar ones (W/O). Solubility, however, is a fairly blunt tool for these types of solutions. Instead, due to their dual nature, surfactant molecules have been described in terms of their hydrophilic and lipophilic character, often reported in a HLB number.

It was generally thought that to form an emulsion, a surfactant needs to have a hydrophilic-lipophilic balance which corresponds to the continuous phase. This means that surfactants with greater water affinity and high HLB are suitable for oil-in-water emulsions and low HLB surfactants are suitable for water-in-oil [1, 6, 97, 102].

This all sounds relatively straightforward, but there are considerable limitations to HLB that are worth summarizing here. Starting simply, cosmetic formulation and other chemical literature often give values of 1–4 for W/O emulsifiers and 10–13 for O/W, but these don't align with those given in the conservation literature [6]. While this could be the result of the different evaluation parameters for what a desirable result looks like, the variation is worth noting [3, 6, 100].

Furthermore, there is also significant ambiguity in how the HLB values are determined. For example, the Griffin scale is based on a simple ratio of the molecular mass of the hydrophilic part of the molecule to the total mass [103]. Davies HLB, on the other hand, is calculated by tallying up values based on the number and type of chemical groups found in the molecule. This is thought to provide a more accurate representation of behaviour as it accounts for the different polarity of different molecular groups [104]. However, the manufacturer supplied literature rarely, and in my experience never, states which system is being employed.

Finally, and most importantly, while HLB can point towards overall trends in behaviour, it is not a de facto predictor of solubility. This is worsened in multi-component solutions. The more complex a solution becomes, the poorer the predictive power of HLB seems to be. This is at least partially because HLB seeks to describe the character of the surfactant alone, independent of other solution components. For this reason, precise emulsion formulation often relies on information from experimental phase behaviour studies, a technique that is well outside the grasp for the average conservator [105].

HYDROPHILIC-LIPOPHILIC DIFFERENCE (HLD)

While the limitations of HLB can frustrate emulsion formation in the conservation laboratory, in the context of large-scale cosmetic and pharmaceutical production and petrochemical extraction, such misfires can result in extraordinarily expensive failed formulations. As a result, several systems have developed in the intervening decades, but current academic emphasis seems to have coalesced on the multi-factor hydrophilic-lipophilic difference (HLD) described by Salager et al [105–108].

Unlike HLB, which attempts to characterise the surfactant in isolation, HLD seeks to balance the interplay of four solution variables: surfactant properties, salinity, temperature, and the alkane character of the oil used [109]. The relationship between these four parameters is modelled as a balancing act, with each component nudging the system one way or another. When the system balances out so that HLD = 0 the hydrophilic and lipophilic character is in balance and thus a microemulsion will form [109]. Before we leap into formulation, as ever, a few terms must be introduced before the discussion can turn to practical matters.

As HLD is ultimately a mathematical expression, the four variables (surfactant character, salinity, temperature, and 'oiliness') must each be represented as a numerical value. Temperature is simply the temperature of the solution in degrees Celsius. The salinity is measured in g/100 mL, but as chlorides are generally Very Bad News for many heritage materials, this will rarely read anything other than zero. The other two, surfactant character and oiliness, will require a little more introduction.

**Characteristic Curvature (Cc):** This is an experimentally determined value unique to each surfactant type which mathematically describes the hydrophilic and lipophilic character of the molecule. Cc values range either side of zero, with a strongly hydrophilic surfactant yielding Cc of -2.5 to a strongly lipophilic 2.5 [109–111]. Unfortunately, as this value is experimentally determined it can be difficult to track down. This is changing, but collated lists are available through Professor Abbott's website and others are listed below [109].

**Effective Alkane Carbon Number (EACN):** The key word in this acronym is 'effective'. The EACN describes the *behaviour* of a compound as compared against simple straight chain hydrocarbon (alkanes) [109]. For example, a simple chain of hexane contains six carbon atoms; its EACN number is 6. The ring-shaped cyclohexane, despite having the same number of carbons (6), behaves more like propane, which has only three carbons. The EACN for cyclohexane is therefore 3 [109]. The EACN value for some conservation solvents can be found online, or calculated using Hansen solubility parameters on the Pirika EACN Estimator website [109, 112].

**Interfacial Tension (IFT)**: One of the benefits of microemulsions, is that the dispersed phase isn't fenced in by a barrier of surfactant molecules. This is because the surface tension between oil and water phases of a true microemulsion is super low. This means that your dispersed phase has almost no barrier between it and the surface, and the cleaning is therefore much more efficient than a classic emulsion. This is because in a standard emulsion, the dispersed phase is protected by a solid wall of surfactant with a (relatively) high IFT.

**Scan**: Finally, rather than looking to one surfactant for all your solubility needs, the HLD system exploits combinations of surfactants to form microemulsion systems. These combinations are worked out in a systematic test known as a scan, the process of which is outlined on the following pages.

Please note that any attempt to outline this entire system of thought within a few short paragraphs is doomed to over-simplify, perhaps to the point of misrepresentation. There is a lot of clever maths and physics behind these ideas, and interested parties are strongly encouraged to see the resources developed by Professor Steven Abbott. The website, book, and associated apps manage to remain comprehensible despite the complexity of the subject matter.

However, as this change is in the process of being adopted, Cc and EACN values are simply not, at the time of writing, readily available. HLB is therefore included in technical data sections, as it is still strongly featured in much of the conservation literature. This is done in full knowledge of its many limitations

*Key Ideas*

- 'O', 'oil', and 'lipid' are used loosely to refer to non-polar solvents while 'W' and 'water' refer to polar.
- In microemulsions, the concentration of surfactant is often much higher than in other cleaning systems. Clearance, usually with a continuous phase solvent, should therefore be carefully considered prior to treatment.
- The continuous phase is the majority component of the emulsion mixture. The 'oil' in the W/O and the water in the O/W.
- The dispersed phase is the minority component of the emulsion mixture. The water in the W/O and the 'oil' in the O/W.
- Because the surfactant is mostly dissolved within the continuous phase, it can be cleared through rinsing with that phase. The surfactant used in an oil in water (o/w) emulsion is water soluble and can be cleared with water. The surfactant used in a water in oil (W/O) emulsion is 'oil' soluble and can be cleared with the 'oily' solvent.
- Hydrophilic surfactants have negative Cc values, while hydrophobic have positive. This means that surfactants like low Cc Tween 20 will be water soluble, while higher Cc Span 20 will be soluble in non-polar solvents.

***Preparation: Microemulsion Testing***

*Techniques and Tips*

Given the breadth of materials involved, there is a near infinite number of possible combinations of solvent, additive, and any of the ever-changing array of surfactants. This inherent complexity and inconsistency of materials confounds any attempt towards concise summary or collation of specific processes or ratios.

However, as with many sections in this book, the following information is provided as an introduction and starting off point for further exploration of contemporaneous materials. For, at the end of the day, even in the most In order to determine which solvents and surfactants will form true microemulsions, despite the sophistication of the HLD system, there is simply no replacement for direct testing.

#### SCAN

A 'scan' is simply a methodical test of two surfactants and two solvents combined in a methodical way. To summarise, each test contains a different ratio of the two surfactants, but the solvent mixture remains constant. This allows the phase changes to be clearly seen and evaluated. Table 4.11 includes example mixtures.

The concentrated surfactants can be highly viscous and difficult to decant in a controlled manner. Rather than measuring out both surfactants into the same small vial, where it is challenging to remove any excess, they can instead be decanted separately and dissolved before being combined. Please note that the surfactant must be fully dissolved for this to work.

#### SURFACTANT CHOICE

Although ethoxylates such as Ecosurf EH6 are seemingly attractive, they have some inherent ambiguity built in. Like all industrial surfactants, they are complex blends of different chain lengths and ethoxylate groups. It happens to be the case that the Cc values of dilute ethoxylates are higher than when they are concentrated. This means that a scan done at relatively high surfactant concentrations will be misleading when used at low concentrations. This doesn't eliminate their usefulness, but just complicates an already complex process.

*Table 4.11* Details of two scans

| *Test* | *APG 12-14* | *Deionised water* | *Span 20* | *White spirit* |
|---|---|---|---|---|
| 1 | 200 mg | 2 mL | 0 | 2 mL |
| 2 | 150 | 2 mL | 50 | 2 mL |
| 3 | 100 | 2 mL | 100 | 2 mL |
| 4 | 50 | 2 mL | 150 | 2 mL |
| 5 | 0 | 2 mL | 200 | 2 mL |
| *Test* | *APG 12-14* | *Xylene* | *Tween 20* | *Deionised water* |
| 6 | 200 mg | 2 mL | 0 | 2 mL |
| 7 | 150 | 2 mL | 50 | 2 mL |
| 8 | 100 | 2 mL | 100 | 2 mL |
| 9 | 50 | 2 mL | 150 | 2 mL |
| 10 | 0 | 2 mL | 200 | 2 mL |

Please note: none of these combinations produced a microemulsion; they are instead illustrative of the scan process.

### ***Preparation: (Micro)Emulsions***

#### *Techniques and Tips*

The emulsification character (whether micro, nano, or regular) is a quality of a solution, and not a preparation technique. As a result, the only difference in preparation is how much agitation it requires. Practically speaking, this really means that although there are circumstances where a true microemulsion is necessary, in others, it's really a question of shaking it until you make it.

##### MIXING

Emulsions are prepared as two separate solutions, each phase mixed with any additional components, before the two are then combined. The surfactant is typically dissolved into liquid that will form the dispersed phase. If the amount of surfactant required in known, this method allows for even dispersion of all components. However, surfactant can also be added slowly until the emulsification forms although this method will result in surfactant-heavy solutions that can prove harder to clear.

Unless aiming for a foamed emulsion (an area of current exploration in textile conservation), avoid high torque or aerating mixing methods, as this can cause the system to become frothy and inconsistent.

##### CLEARANCE

The solvent which makes up the continuous phase, or one of very similar character, is most often selected to clear away a given emulsion. However, please note that systems which use a high ratio of surfactant to solvent will likely require multiple clearance steps.

#### *Method: Known Parts*

##### MATERIALS

Graduated cylinder
Pipettes
Balance (if using solid additives)
Lidded container
Magnetic stirrer, frother, or glass stirring rod

##### INSTRUCTIONS

1. Measure out full quantity of water. This is regardless of whether water is used as the continuous or the dispersed phase.
2. Add buffering or chelation agents and adjust pH to desired level (see the self-titled sections in Chapter 4 for additional information).
3. Measure out full amount of non-polar solvent or 'oil' component.

4. Measure out surfactant and add to the dispersed phase.
5. Mix surfactant with the dispersed phase using a glass stirring rod. Again, avoid whisking air into the solution (unless foam is desired).
6. Pour the continuous phase into the lidded container and add the dispersed phase.
7. Mix gently with a stirring rod or by swirling the container.
8. The two phases should mix without vigorous stirring and should form a clear or slightly milky solution.

*Method: Unknown Parts (Mayonnaise Style)*

MATERIALS

Graduated cylinder
Pipettes
Balance (if using solid additives)
Lidded container
Magnetic stirrer, frother, or glass stirring rod

INSTRUCTIONS

1. Measure out full quantity of water. This is regardless of whether water is used as the continuous or the dispersed phase.
2. Add buffering or chelation agents to the water and adjust pH to desired level (see the self-titled sections in Chapter 4 for additional information).
3. Measure out full amount of solvent or 'oil' component.
4. Measure out and record a small quantity of surfactant.
5. Add the surfactant to the dispersed phase.
6. Mix surfactant with the dispersed phase using a glass stirring rod. Again, avoid whisking air into the solution (unless foam is desired).
7. Place the lidded container on the magnetic stirrer and add the continuous phase. Set the stirrer on a medium-low setting and add the dispersed phase.
8. Stop the stirrer and check the solution. If it separates immediately or large droplets are visible, restart the stirrer and add more surfactant.
9. Repeat step 8 until the solution is homogenous and clear or slightly milky in appearance.
10. Note the amount of surfactant used.

*Cleanup*

Surfactant disposal methods should be followed for emulsion systems, with care taken to avoid release of hazardous materials into the water systems. Further steps may also be required to neutralise pH or avoid release of hazardous chelation agents.

See manufacturer or supplier information for correct disposal methods.

***Technical Data***

*Table 4.12* Some examples of emulsions from the conservation literature

| *Solvent* | *mL* | *Surfactant* | *HLB* | *mL* | *Aqueous solution* | *mL* | *Citation* |
|---|---|---|---|---|---|---|---|
| Xylene | 60 | Triton x-100 | 13.4 | 30 | 100 mL<br>1.36 g $KH_2PO_4$<br>1.1 g diammonium citrate<br>6.0 pH with NaOH | 25 | [3] |
| Xylene | 50 | Triton x-100 | 13.4 | 20 | 1 part TEA: 30 parts water<br>8.5 pH HCl | 30 | [3] |
| Cyclomethicone | 50 | EcoSurf EH3 | 7.9 | 25 | EDTA and Buffer | 30 | [113] |
| mineral spirits | | Brij 35 | 16.9 | | 2% TEA in water | | [3] |
| Decamethylcyclopentasiloxane | 25 | EcoSurf EH3 | 7.9 | 75 | Water<br>Cleared with octamethyltrisoloxane | 5 | [114] |
| decamethylcyclopentasiloxane | 25 | EcoSurf EH3 | 7.9 | 75 | 100mL water<br>0.5g EDTA<br>0.5g citric acid<br>6.5 pH with NaOH<br>Cleared with octamethyltrisoloxane | 25 | [114] |

*Table 4.13* Data for a selection of currently available or frequently referenced surfactants used as emulsifiers

| *Commercial name* | *Producer/ supplier* | *Soluble in* | *Type* | *HLB* | *Density g/mL** | *pH*** | *Source* |
|---|---|---|---|---|---|---|---|
| APG 12-14 | Various | Water, alcohol | Non-ionic | 12 | 1.16 | 11.5-12.5 | [13] |
| Brij-35 | Croda | Water, alcohol | Non-ionic | 16.9 | 1.05 | 5–7 | [14] |
| Brij S 100 | Croda | Water, alcohol | Non-ionic | 18.8 | Not reported | Not reported | [15] |
| Ecosurf EH-3 | DOW | Water, polar organic and chlorinated solvents. | Non-ionic | 7.9 | 0.9651 | 6.2 | [20, 21] |
| Ecosurf EH-6 | DOW | Water | Non-ionic | 10.8 | 0.9897 | 6.2 | [21, 22] |
| Ecosurf EH-9 | DOW | Water | Non-ionic | 12.5 | 1.0069 | 6.2 | [115] |
| Span 20 | Croda | Xylene, mineral oil, other non-polar. **INsoluble** in water | Non-ionic | 8.6 | 1.058 | N/A | [24] |
| Tergitol 15-S-3 | DOW | Polar organic and chlorinated solvents. **INsoluble** in water | Non-ionic | 8.0 | 0.918 | N/A | [25, 26] |
| Tergitol 15-S-9 | DOW | Water, polar organic and chlorinated solvents. | Non-ionic | 13.3 | 1.006 | 7.1 | [26, 27] |
| Triton X-100* | DOW | Water, alcohols, organic and chlorinated solvents | Non-ionic | 13.4 | 1.061 | 6 | [10, 28] |
| Tween 20 | Croda | Water, polar solvents likely | Non-ionic | 16.7 | 1.1 | 6–8 | [3, 29] |
| Tween 80 | Croda | Water, polar solvents likely | Non-ionic | 15.0 | 1.06 | 5–7 | [3, 30–32] |

* see TDS for details.
** 1–10% aqueous solutions reported, see individual technical data sheets for more details.

## References

[1] Y. Nakama, "Surfactants," in *Cosmetic Science and Technology: Theoretical Principles and Applications*, Elsevier, 2017, pp. 231–244. doi: 10.1016/B978-0-12-802005-0.00015-X
[2] K. Kosswig, "Surfactants," in *Ullmann's Encyclopedia of Industrial Chemistry*, Weinheim, Germany: Wiley-VCH Verlag GmbH & Co. KGaA, 2000. doi: 10.1002/14356007.a25_747
[3] R. Wolbers, *Cleaning Painted Surfaces: Aqueous Methods*. London: Archetype Publications, 2000.
[4] Á. Tímár-Balázsy and D. Eastop, *Chemical Principles of Textile Conservation*. Butterworth-Heinemann series in conservation and museology. Oxford [England]; Boston: Butterworth-Heinemann, 1998.
[5] S. Rivers and N. Umney, *Conservation of Furniture*. Abingdon: Routledge, 2013.
[6] Y. Yamashita, R. Miyahara, and K. Sakamoto, "Emulsion and emulsification technology," in *Cosmetic Science and Technology: Theoretical Principles and Applications*, Elsevier Inc., 2017, pp. 489–506. doi: 10.1016/B978-0-12-802005-0.00028-8
[7] D. Stulik and R. Wolbers, "Project Outcome, Spin-offs, and Future Research Needs," in *Solvent Gels for the Cleaning of Works of Art: The Residue Question*, in Research in Conservation, Los Angeles: Getty Publishing, 2004.
[8] J. A. Fields, A. Wingham, F. Hartog, and V. Daniels, "Finding Substitute Surfactants for Synperonic N," *Journal of the American Institute for Conservation*, vol. 43, no. 1, pp. 55–73, 2004, doi: 10.1179/019713604806112579
[9] M. Palmer and H. Hatley, "The role of surfactants in wastewater treatment: Impact, removal and future techniques: A critical review," *Water Research*, vol. 147, pp. 60–72, 2018, doi: 10.1016/j.watres.2018.09.039.
[10] C. Stavroudis, "Health & Safety: The Topic is Estrogenic Chemicals," *WAAC Newsletter*, vol. 17, no. 1, 1995. Accessed: . [Online]. Available: https://cool.culturalheritage.org/waac/wn/wn17/wn17-1/wn17-103.html
[11] L. Pomerantz, "Treatment and mounting of Ceres, a large mural painting on fabric by John Norton," *Journal of the American Institute of Conservation*, vol. 24, no. 1,1984, pp. 14–22. Accessed: Jul. 29, 2023. [Online]. Available: https://cool.culturalheritage.org/jaic/articles/jaic24-01-002.html
[12] Solvay, "Aerosol OT-75 Surfactant," 2023.
[13] "Lauryl glucoside | 110615-47-9," ChemicalBook. Accessed: Jun. 12, 2024. [Online]. Available: https://www.chemicalbook.com/ChemicalProductProperty_EN_CB2965619.htm
[14] ThermoFischer, "Brij™-35," ThermoFischer. Accessed: Aug. 31, 2023. [Online]. Available: https://www.thermofisher.com/order/catalog/product/20150
[15] Croda Chemical Company, "Brij™ S100." Accessed: Aug. 31, 2023. [Online]. Available: https://www.crodacropcare.com/en-gb/product-finder/product/10-brij_1_s100
[16] J. Hackett, "Publication Contribution - Conservator's Cookbook," [Personal Communication] Aug. 03, 2024.
[17] "Dehypon LS 45," In Situ Conservation. Accessed: Aug. 31, 2023. [Online]. Available: https://www.insituconservation.com/en/products/soaps/dehypon_ls45
[18] M. Sato and A. Quye, "Detergency evaluation of non-ionic surfactant Dehypon® LS54 for textile conservation wet cleaning," *Journal of the Institute of Conservation*, vol. 42, no. 1, pp. 3–17, 2019, doi: 10.1080/19455224.2018.1556719
[19] BASF, "Dehypon LS 54 Technical Data Sheet," 2015. [Online]. Available: https://worldaccount.basf.com
[20] Dow Chemical Company, "Ecosurf™ EH-3 Surfactant," Technical Data Sheet. [Online]. Available: http://www.dow.com/surfactants

[21] C. Stavroudis, "More from CAPS3: Surfactants, silicone-based solvents, and microemulsions," *WAAC Newsletter*, vol. 34, no. 3, pp. 24–27, 2012. Accessed: Jul. 29, 2023. [Online]. Available: https://cool.culturalheritage.org/waac/wn/wn34/wn34-3/wn34-306.pdf
[22] Dow Chemical Company, "Ecosurf™ EH-6 Surfactant," Technical Data Sheet. [Online]. Available: http://www.dow.com/surfactants
[23] Clariant Chemical Company, "Hostapon ® TPHC Mild, Anionic surfactant for the cosmetic industry," 2020. [Online]. Available: www.ics.clariant.comwww.clariant.com
[24] Merck, "Span® 20 CAS 1338-39-2 | 840119." Accessed: Jun. 11, 2024. [Online]. Available: https://www.merckmillipore.com/GB/en/product/Span-20,MDA_CHEM-840119
[25] Dow Chemical Company, "Tergitol™ 15-S-3 Surfactant," Technical Data Sheet. [Online]. Available: http://www.dow.com/surfactants
[26] G. St-Laurent, "The Care and Handling of Recorded Sound Materials." Accessed: Jul. 26, 2023. [Online]. Available: https://cool.culturalheritage.org/byauth/st-laurent/care.html
[27] Dow Chemical Company, "Tergitol™ 15-S-9 Surfactant," Technical Data Sheet. [Online]. Available: http://www.dow.com/surfactants
[28] Dow Chemical Company, "Triton X-100 Surfactant," Technical Data Sheet. [Online]. Available: http://www.dow.com/surfactants
[29] Croda Chemical Company, "Tween™ 20 Surfactant," Technical Data Sheet. [Online]. Available: https://www.crodacropcare.com/en-gb/product-finder/product/248-tween_1_20
[30] Croda Chemical Company, "Tween™ 80 Surfactant," Technical Data Sheet. [Online]. Available: https://www.crodahomecare.com/en-gb/product-finder/product/4308-eco_1_tween_1_80
[31] S. Tiwari, C. Mall, and P. P. Solanki, "CMC studies of CTAB, SLS & tween 80 by spectral and conductivity methodology to explore its potential in photogalvanic cell," *Surfaces and Interfaces*, vol. 18, p. 100427, 2020, doi: 10.1016/j.surfin.2019.100427
[32] S. Chatterjee, A. Jain, and S. De, "Effect of different operating conditions in cloud point assisted extraction of thymol from Ajwain (Trachyspermum Ammi L.) seeds and recovery using solvent," *Journal of Food Science Technology*, vol. 54, no. 13, pp. 4353–4361, 2017, doi: 10.1007/s13197-017-2906-z
[33] A. Thorn, "The impact of disodium EDTA on stone," in *ICOM Committee for Conservation tenth triennial meeting*, Washington, DC, Aug. 1993, pp. 357–363.
[34] Merck, "Sigma-Aldrich Millipore Buffer Calculator." [Online]. Available: https://www.sigmaaldrich.com/GB/en/support/calculators-and-apps/buffer-calculator
[35] N. A. Lange and J. A. Dean, *Lange's Handbook of Chemistry*, 15th ed. New York: McGraw-Hill, 1998.
[36] Dow Chemical Company, "Chelation chemistry General concepts of the chemistry of chelation," 113-01388-01-0721 S2D, 2021. [Online]. Available: https://www.dow.com/content/dam/dcc/documents/en-us/app-tech-guide/113/113-01388-01-chelation-chemistry-general-concepts-of-the-chemistry-of-chelation.pdf
[37] W. D. Richey, "Chelating agents – a review," *Studies in Conservation*, vol. 20, no. sup1, pp. 229–234, 1975, doi: 10.1179/sic.1975.s1.038
[38] H. J. Plenderleith, *The Conservation of Antiquities and Works of Art: Treatment, Repair, and Restoration*. London: Oxford University Press, 1956.
[39] F. Basolo, Y. T. Chen, and R. K. Murmann, "Steric effects and the stability of complex compounds. IV. the chelating tendencies of c-substituted ethylenediamines with copper(II) and nickel(II) ions1,2," *Journal of American Chemical Society*, vol. 76, no. 4, pp. 956–959, 1954, doi: 10.1021/ja01633a002.

[40] NIST Chemistry, *National Institute for Standards and Technology (NIST) Chemistry WebBook*. National Institute of Standards and Technology, 2023. Accessed: Sep. 01, 2023. [Online]. Available: https://webbook.nist.gov/cgi/cbook.cgi?Name=EDTA&Units=SI
[41] T. Higgins, "Publication Contribution - Conservator's Cookbook," [Personal Communication] May 29, 2024.
[42] C. Gervais, C. A. Grissom, N. Little, and M. J. Wachowiak, "Cleaning marble with ammonium citrate," *Studies in Conservation*, vol. 55, no. 3, pp. 164–176, 2010, doi: 10.1179/sic.2010.55.3.164
[43] R. Morrison, A. Bagley-Young, A. Burnstock, K. J. Van Den Berg, and H. Van Keulen, "An investigation of parameters for the use of citrate solutions for surface cleaning unvarnished paintings," *Studies in Conservation*, vol. 52, no. 4, pp. 255–270, 2007, doi: 10.1179/sic.2007.52.4.255
[44] B. Prestowitz, C. Theodorakopoulos, and J. Colbourne, "Preliminary investigation into the use of diethylenetriaminepentaacetic acid and ethylenediaminetetraacetic acid to treat iron induced foxing in paper objects," *The Book and Paper Group Annual*, vol. 35, pp. 163–170, 2016.
[45] M. Brockman and E. Farek, "Investigation into reduction of foxing stains in paper," presented at the Association of North American Graduate Programs in Conservation, Queens University, Kingston, Ontario, 2018. Accessed: Sep. 01, 2023. [Online]. Available: https://resources.culturalheritage.org/wp-content/uploads/sites/11/2019/04/2018Foxing-Investigation-ANAGPIC-Paper.pdf
[46] A. Bartoletti *et al.*, "Facilitating the conservation treatment of Eva Hesse's Addendum through practice-based research, including a comparative evaluation of novel cleaning systems," *Heritage Science*, vol. 8, no. 1, p. 35, 2020, doi: 10.1186/s40494-020-00378-z
[47] A. Moskalik-Detalle, J. Assoun, F. Joseph, M.-L. Martiny, and M. Monfort, "Conservation of murals by Eugène Delacroix at Saint Sulpice, Paris," in *Gels in the Conservation of Art*, L. V. Angelova, B. Ormsby, J. Townsend, and R. Wolbers, Eds., London: Archetype, 2017, pp. 200–208.
[48] D. G. Suryawanshi and S. K. Bisaria, "Removing metallic stains from paper objects using chelating agent EDTA," *Restaurator*, vol. 26, no. 4, 2005, doi: 10.1515/REST.2005.276
[49] Richmond Conservation Studio, "Washing grey skies away… with a little HEDTA…. Nice ring to it eh?," Facebook Post. Accessed: Aug. 22, 2023. [Online]. Available: https://www.facebook.com/RichmondConservationStudio/videos/washing-grey-skies-away-with-a-little-hedta-nice-ring-to-it-eh-%EF%B8%8Fpaintingconserva/3954604787921751/
[50] K. Berrett, V. Naude, and R. Wolbers, "A new method for cleaning marble," *Objects Specialty Group Postprints*, vol. 14, 2008. Accessed: Jul. 01, 2023. [Online]. Available: https://resources.culturalheritage.org/osg-postprints/wp-content/uploads/sites/8/2015/02/osg014-12.pdf
[51] C. Stavroudis, T. Doherty, and R. Wolbers, "A New Approach to Cleaning I: Using Mixtures of Concentrated Stock Solutions and a Database to Arrive at an Optimal Aqueous Cleaning System," *WAAC Newsletter*, vol. 27, no. 2, 2005. Accessed: Aug, 2023. [Online]. Available: https://cool.culturalheritage.org/waac/wn/wn27/wn27-2/wn27-205.pdf
[52] T. Stambolov, "Notes on the Removal of Iron Stains from Calcareous Stone," *Studies in Conservation*, vol. 13, no. 1, pp. 45–47, 1968, doi: 10.2307/1505388.
[53] L. Carlyle, J. Townsend, and S. Hackney, "Triammonium citrate an investigation into its application for surface cleaning," in *Dirt and Pictures Separated: Papers Given at a Conference Held Jointly by UKIC and the Tate Gallery, January 1990*, S. Hackney, J. Townsend, and N. Eastaugh, Eds., London (United Kingdom): United Kingdom Institute of Conservation, pp. 44–48. Accessed: Jan. 12, 2024. [Online].

Available: https://www.researchgate.net/profile/Stephen-Hackney/publication/312368609_'Triammonium_citrate_an_investigation_into_its_application_for_surface_cleaning'/links/58907dea92851c9794c677ac/Triammonium-citrate-an-investigation-into-its-application-for-surface-cleaning.pdf

[54] A. Dwan, "Ammonium citrates for stain removal in paper," *WAAC Newsletter*, vol. 37, no. 3, 2015.

[55] S. Adler and L. Eaton, "Chelating Agents in Wet Cleaning Systems," *AIC Textile Specialty Group Postprints*, pp. 69–78, 1995.

[56] T. Markevičius, T. Syversen, E. Chan, N. Olsson, C. Skov Hilby, and R. Šimaitė, "Cold, warm, warmer: use of precision heat transfer in the optimization of hydrolytic enzyme and hydrogel cleaning systems," in *Gels in the Conservation of Art*, L. V. Angelova, B. Ormsby, J. Townsend, and R. Wolbers, Eds., London: Archetype, 2017, pp. 67–72.

[57] M. El-Gohary and M. Saad, "Using some traditional and nanomaterials in the cleaning of ancient Egyptian pottery artifacts (jar and shallow plate) from Athribis," *Archaeological and Anthropological Sciences*, vol. 12, no. 12, 2020, doi: 10.1007/s12520-020-01235-7

[58] M. Sullivan, S. Brogdon-Grantham, and K. Taira, "New approaches to cleaning works of art on paper and photographs," in *Student Papers & Posters*, Buffalo State College, 2014. [Online]. Available: https://faic.wpenginepowered.com/anagpic-student-papers/wp-content/uploads/sites/11/2016/02/anagpic2014_sullivan_etal_paper.pdf

[59] L. Mina, "Foxy underpants: or the use of chelators and enzymes to reduce foxing stains on early nineteenth century men's linen underpants," *Journal of the American Institute for Conservation*, vol. 59, no. 1, pp. 3–17, 2020, doi: 10.1080/01971360.2019.1674604

[60] W. Aehle, R. N. Perham, G. Michal, A. J. Caddow, and B. Concoby, "Enzymes, 1. general," in *Ullmann's Encyclopedia of Industrial Chemistry*, Wiley-VCH Verlag GmbH & Co. KGaA, 2008. doi: 10.1002/14356007.a09_341.pub3

[61] N. Barbabietola *et al.*, "A safe microbe-based procedure for a gentle removal of aged animal glues from ancient paper," *International Biodeterioration & Biodegradation*, vol. 109, pp. 53–60, 2016, doi: 10.1016/J.IBIOD.2015.12.019

[62] F. Cappitelli *et al.*, "Advantages of using microbial technology over traditional chemical technology in removal of black crusts from stone surfaces of historical monuments," *Applied and Environmental Microbiology Journal*, vol. 73, no. 17, pp. 5671–5675, 2007, doi: 10.1128/AEM.00394-07

[63] P. Bosch Roig, J. L. Regidor Ros, and R. Montes Estellés, "Biocleaning of nitrate alterations on wall paintings by Pseudomonas stutzeri," *International Biodeterioration & Biodegradation*, vol. 84, pp. 266–274, 2013, doi: 10.1016/j.ibiod.2012.09.009

[64] P. Bosch-Roig and G. Ranalli, "The safety of biocleaning technologies for cultural heritage," *Frontiers in Microbiology*, vol. 5, no. Apr, 2014, doi: 10.3389/fmicb.2014.00155

[65] P. Bosch-Roig, G. Lustrato, E. Zanardini, and G. Ranalli, "Biocleaning of cultural heritage stone surfaces and frescoes: which delivery system can be the most appropriate?," *Annals of Microbiology*, vol. 65, no. 3, pp. 1227–1241, 2015, doi: 10.1007/s13213-014-0938-4

[66] P. Bosch-Roig, J. L. Regidor Ros, M. P. Soriano Sancho, R. Montes Estellés, and P. Roig Picazo, "Biocleaning of wall paintings on uneven surfaces with warm agar gels," in *Gels in the Conservation of Art*, L. V. Angelova, B. Ormsby, J. Townsend, and R. Wolbers, Eds., London: Archetype, 2017, pp. 119–121.

[67] G. Ranalli and E. Zanardini, "Biocleaning on Cultural Heritage: new frontiers of microbial biotechnologies," *Journal of Applied Microbiology*, vol. 131, no. 2, pp. 583–603, 2021, doi: 10.1111/jam.14993

[68] P. C. DeSantis, "Some observations on the use of enzymes in paper conservation," *Journal of the American Institute for Conservation*, vol. 23, no. 1, pp. 7–27, 1983.

[69] H. M. Erickson, "Usage recommendations for alpha-amylases: Maximizing enzyme activity while minimizing enzyme-artifact binding residues," *The Book and Paper Group Annual*, pp. 24–33, 1992.

[70] H. Taniguchi and Y. Honnda, "Amylases," in *Encyclopedia of Microbiology*, 3rd ed., M. Schaechter, Ed., Oxford: Academic Press, 2009, pp. 159–173. doi: 10.1016/B978-012373944-5.00130-9

[71] P. Cremonesi and A. Casoli, "Enzymes as tools for conservation of works of art," *Journal of Cultural Heritage*, vol. 50, pp. 73–87, 2021, doi: 10.1016/j.culher.2021.06.005

[72] F. Palla, "Bioactive molecules: innovative contributions of biotechnology to the restoration of cultural heritage," *Conservation Science in Cultural Heritage*, vol. 13, pp. 369–373, 2013, doi: 10.6092/ISSN.1973-9494/4191

[73] M. Abdel-Nasser, G. Abdel-Maksoud, M. S. Abdel-Aziz, S. S. Darwish, A. A. Hamed, and A. M. Youssef, "Evaluation of the efficiency of nanoparticles for increasing α-amylase enzyme activity for removing starch stain from paper artifacts," *Journal of Cultural Heritage*, vol. 53, pp. 14–23, 2022, doi: 10.1016/j.culher.2021.11.004

[74] F. Cappitelli *et al.*, "Advantages of using microbial technology over traditional chemical technology in removal of black crusts from stone surfaces of historical monuments," *Applied and Environmental Microbiology Journal*, vol. 73, no. 17, pp. 5671–5675, 2007, doi: 10.1128/AEM.00394-07

[75] F. Gherardi, L. Turyanska, E. Ferrari, N. Weston, M. W. Fay, and B. J. Colston, "Immobilized enzymes on gold nanoparticles: from enhanced stability to cleaning of heritage textiles," *ACS Applied Bio Materials*, vol. 2, no. 11, pp. 5136–5143, 2019, doi: 10.1021/acsabm.9b00802.

[76] C. Skov Hilby, "Publication Contribution - Conservator's Cookbook," [Personal Communication] May 29, 2024.

[77] Y. van Dyke, "Agarose-enzyme gels in paper conservation," in *Gels in the Conservation of Art*, L. V. Angelova, B. Ormsby, J. Townsend, and R. Wolbers, Eds., London: Archetype Publications, 2017, pp. 101–106.

[78] S. Glenn, J. Hackett, E.-A. Haldane, and S. Im, "Borrowing from the neighbours: using the technology of other disciplines to treat difficult textile conservation problems," in *Learning Curve: Education, Experience, Reflection. Postprints*, A. Fairhurst, Ed., Birkbeck College, London: ICON, Apr. 2015, pp. 33–42. [Online]. Available: https://www.icon.org.uk/static/dd33714f-f5b2-4f69-8b1eb6d866d4a8b0/2015learningcurveforum.pdf

[79] Y. van Dyke, "Practical applications of protease enzymes in paper conservation," *The Book and Paper Group Annual*, vol. 23, pp. 93–107, 2004.

[80] S. Barbisan and A.-L. Dupont, "Local cleaning of tidelines on paper using rigid gels: the influence of pH and conductivity," in *Gels in the Conservation of Art*, London, UK: Archetype Publications Ltd., 2017.

[81] A. Bazemore, "Chelating soluble iron (II) from iron gall ink using calcium phytate in agar gel," in *Gels in the Conservation of Art*, London, UK: Archetype Publications Ltd., 2017.

[82] N. Bonelli, G. Poggi, D. Chelazzi, R. Giorgi, and P. Baglioni, "Poly(vinyl alcohol)/poly(vinyl pyrrolidone) hydrogels for the cleaning of art," *Journal of Colloid and Interface Science*, vol. 536, pp. 339–348, 2019, doi: 10.1016/j.jcis.2018.10.025

[83] M. Tortora, F. Gherardi, E. Ferrari, and B. Colston, "Biocleaning of starch glues from textiles by means of α-amylase-based treatments," *Applied Microbiology and Biotechnology*, vol. 104, no. 12, pp. 5361–5370, 2020, doi: 10.1007/s00253-020-10625-9

[84] T. M. Andrews, W. W. Andrews, and C. Baker, "An investigation into the removal of enzymes from paper following conservation treatment," *Journal of the American Institute for Conservation*, vol. 31, no. 3, pp. 313–323, 1992, doi: 10.1179/019713692806066547

[85] H. E. Ahmed, S. S. Gremos, and F. N. Kolisis, "Enzymatic removal of the oily dirt from a Coptic tunic using the enzyme lipase," *Journal of Textile and Apparel, Technology and Management*, vol. 6, no. 3, 2010, Accessed: Jul. 15, 2023. [Online]. Available: https://jtatm.textiles.ncsu.edu/index.php/JTATM/article/view/674

[86] BRENDA, "BRENDA Enzyme Database." Accessed: Sep. 01, 2023. [Online]. Available: https://www.brenda-enzymes.org/index.php

[87] C. Mazzuca *et al.*, "Cleaning of paper artworks: Development of an efficient gel-based material able to remove starch paste," *ACS Applied Materials and Interfaces*, vol. 6, no. 19, pp. 16519–16528, 2014, doi: 10.1021/am504295n

[88] M. Ferrari *et al.*, "Enzymatic laundry for old clothes: immobilized alpha-amylase from Bacillus sp. for the biocleaning of an ancient Coptic tunic," *Applied Microbiology and Biotechnology*, vol. 101, no. 18, pp. 7041–7052, 2017, doi: 10.1007/s00253-017-8437-8

[89] H. E. Ahmed and F. N. Kolisis, "An investigation into the removal of starch paste adhesives from historical textiles by using the enzyme α-amylase," *Journal of Cultural Heritage*, vol. 12, no. 2, pp. 169–179, 2011, doi: 10.1016/j.culher.2010.08.001

[90] N. J. Borring and C. Skov Hilby, "Enzymes in conservation: separating the Triumphal Arch by Albrecht Dürer with α-amylase embedded in agarose gels," in *IADA Congress XIII*, Berlin, 2015. doi: 10.13140/RG.2.1.1505.1609

[91] N. J. Borring, H. Singer, C. Skov Hilby, and L. Klyver, "Dürer Under The Knife! Conservation of The Arch Of Honour Of Maximilian," in *Power and Glory – Albrecht Dürer in the Emperor's Service*, N. J. Borring and H. Kolind Poulsen, Eds., Copenhagen: Statens Museum for Kunst, 2015, pp. 82–92.

[92] R. Bellucci, P. Cremonesi, and G. Pignagnoli, "A preliminary note on the use of enzymes in conservation: the removal of aged acrylic resin coatings with lipase," *Studies in Conservation*, vol. 44, no. 4, 1999, pp. 278–281, doi: 10.1179/sic.1999.44.4.278

[93] J. Guo *et al.*, "Optimization of conditions for the cleaning of overpaint from the stone carvings of the Potala Palace, Tibet using enzyme-based cleaning agents," *Journal of Cultural Heritage*, vol. 59, pp. 83–92, 2023, doi: 10.1016/j.culher.2022.11.009

[94] S. Abdelaal and I. C. A. Sandu, "Assessment of protease in cleaning of bat blood patches from ancient Egyptian wall paintings and surface inscriptions," *International Journal of Conservation Science*, vol. 10, no. 3, pp. 459–474, 2019.

[95] H. E. Ahmed and F. N. Kolisis, "A study on using of protease for removal of animal glue adhesive in textile conservation," *Journal of Applied Polymer Science*, vol. 124, no. 5, pp. 3565–3576, 2012, doi: 10.1002/app.34053

[96] G. Petrella *et al.*, "A new sustainable and innovative work for paper artworks cleaning process: gellan hydrogel combined with hydrolytic enzymes," *International Journal of Conservation Science*, vol. 7, no. speciall, 2016.

[97] C. Solans and M. J. García-Celma, "Microemulsions and nano-emulsions for cosmetic applications," in *Cosmetic Science and Technology: Theoretical Principles and Applications*, Elsevier, 2017, pp. 507–518. doi: 10.1016/B978-0-12-802005-0.00029-X

[98] Y. Yang *et al.*, "An overview of Pickering emulsions: solid-particle materials, classification, morphology, and applications," *Frontiers in Pharmacology*, vol. 8, 2017, doi: 10.3389/fphar.2017.00287

[99] A. Utomo, N. J. Alderman, G. A. Padron, and N. G. Özcan-Taşkın, "A comparative study on emulsification in the presence of a nanoclay (Pickering emulsion) or a surfactant using high intensity mixing," *Chemical Engineering Research and Design*, vol. 197, pp. 617–627, 2023, doi: 10.1016/j.cherd.2023.07.041

[100] R. Wolbers, "Terminology and properties of selected gels," in *Gels in the Conservation of Art*, L. V. Angelova, B. Ormsby, J. Townsend, and R. Wolbers, Eds., London: Archetype, 2017, pp. 381–394.

[101] M. Reger, T. Sekine, and H. Hoffmann, "Pickering emulsions stabilized by amphiphile covered clays," *Colloids and Surfaces A: Physicochemical and Engineering Aspects*, vol. 413, pp. 25–32, 2012, doi: 10.1016/j.colsurfa.2011.12.005

[102] C. Miller, "Solubilization in surfactant systems," in *Handbook of Surface and Colloid Chemistry*, 3rd ed, K. Birdi, Ed., CRC Press, 2008, pp. 415–438. doi: 10.1201/9781420007206.ch9

[103] W. C. Griffin, "Classification of Surface-Active Agents by 'HLB,'" *Journal of the Society of Cosmetic Chemists*, vol. 1, no. 5, pp. 311–326, 1949.

[104] J. T. Davies, "A quantitative kinetic theory of emulsion type, I. Physical chemistry of the emulsifying agent," in *Gas/Liquid and Liquid/Liquid Interface*, 1957, pp. 426–438. [Online]. Available: https://citeseerx.ist.psu.edu/document?repid=rep1&type=pdf&doi=3247eead9a12ee37c1216fd7ab95ebd863823c88

[105] A. Witthayapanyanon, J. H. Harwell, and D. A. Sabatini, "Hydrophilic–lipophilic deviation (HLD) method for characterizing conventional and extended surfactants," *Journal of Colloid and Interface Science*, vol. 325, no. 1, pp. 259–266, 2008, doi: 10.1016/j.jcis.2008.05.061

[106] J.-L. Salager, N. Marquez, A. Graciaa, and J. Lachaise, "Partitioning of ethoxylated octylphenol surfactants in microemulsion−oil−water systems: influence of temperature and relation between partitioning coefficient and physicochemical formulation," *Langmuir*, vol. 16, no. 13, pp. 5534–5539, 2000, doi: 10.1021/la9905517

[107] L. Jin *et al.*, "Predicting microemulsion phase behavior using physics based HLD-NAC equation of state for surfactant flooding," *Journal of Petroleum Science and Engineering*, vol. 151, pp. 213–223, 2017, doi: 10.1016/j.petrol.2016.12.035

[108] J. L. Salager, M. Bourrel, R. S. Schechter, and W. H. Wade, "Mixing rules for optimum phase-behavior formulations of surfactant/oil/water systems," *Society of Petroleum Engineers Journal*, vol. 19, no. 05, pp. 271–278, 1979, doi: 10.2118/7584-PA.

[109] S. Abbott, "Practical Surfactants," Prof Steven Abbott. [Online]. Available: https://www.stevenabbott.co.uk/practical-surfactants/basics.php

[110] A. B. Mejia Jr., "Characteristic curvature evaluation of anionic surfactants for enhanced oil recovery applications," in *Proceedings of the National Conference On Undergraduate Research (NCUR)*, Eastern Washington University Cheney, WA, Apr. 2015. [Online]. Available: https://libjournals.unca.edu/ncur/wp-content/uploads/2021/07/1224-Mejia-FINAL-W-Revisions.pdf

[111] S. Zarate-Muñoz, F. Texeira de Vasconcelos, K. Myint-Myat, J. Minchom, and E. Acosta, "A simplified methodology to measure the characteristic curvature (Cc) of alkyl ethoxylate nonionic surfactants," *Journal of Surfactants and Detergents*, vol. 19, no. 2, pp. 249–263, 2016, doi: 10.1007/s11743-016-1787-x

[112] H. Yamamoto, "Estimation of EACN (equivalent alkane carbon number) with HSP," Pirika Hansen solubility parameters(HSP) application notes. Accessed: May 28, 2024. [Online]. Available: https://pirika.com/NewHP/PirikaE3/EACN.html

[113] Getty Conservation Institute, *Preparing a Dow Mineral Spirits Microemulsion (With Cosurfactants)*, (Oct. 31, 2013). Accessed: Aug. 16, 2023. [Online Video]. Available: https://www.youtube.com/watch?v=SGkf3i7rnDw

[114] J. Hackett, "Cleaning PVC with microemulsions," *V & A Conservation Journal*, no. 62, 2014, [Online]. Available: http://www.vam.ac.uk/content/journals/conservation-journal/autumn-2014-issue-62/cleaning-pvc-with-microemulsions-joanne-hackett/

[115] Dow Chemical Company, "Ecosurf™ EH-9 Surfactant," Technical Data Sheet, 2022. [Online]. Available: https://www.dow.com/documents/119/119-02243-01-ecosurf-eh9-surfactant-tds.pdf?iframe=true

# Appendix: Works Consulted

## 1 – Bench Skills

Ballard, A. J., & Dellago, C. (2012). Toward the mechanism of ionic dissociation in water. *Journal of Physical Chemistry B, 116*(45), 13490–13497, DOI: 10.1021/jp309300b.

Bartoletti, A., Soares, I., Ramos, A. M., Shashoua, Y., Quye, A., Casimiro, T., & Ferreira, J. L. (2023). Assessing the impact and suitability of dense carbon dioxide as a green solvent for the treatment of PMMA of historical value. *Polymers, 15*(3). DOI: 10.3390/polym15030566

Bell, S. (2001). *A Beginner's Guide to Uncertainty of Measurement* (Good Practice Guide No. 11). National Physical Laboratory. [online] Available at: https://www.npl.co.uk/gpgs/beginners-guide-measurement-uncertainty-gpg11 [Accessed 13 Apr. 2024].

Cammidge, E. M. (1991). Benzene and turpentine: The pre-history of drycleaning. *Ambix: The Journal of the Society for the Study of Alchemy and Early Chemistry, 38*(2), 79–84.

Chaillie, L. (2024, April 17). *Solvent Use in 2024 - Survey Results* [Discussion]. Global Conservation Forum (ConsDistList). [online] Available at: https://community.culturalheritage.org/discussion/solvent-use-in-2024-survey-results?ReturnUrl=%2fcommunities%2fcommunity-home%2fdigestviewer%3fListKey%3d757a8f16-505f-4323-8e74-e376757aa9f7%26CommunityKey%3dea3d002c-9fc3-4446-b7d2-c308f5faed13 [Accessed 25 Apr. 2024]

Davidge, J., Thomas, C. P., & Williams, D. R. (2001). Conditional formation constants or chemical speciation data? *Chemical Speciation & Bioavailability, 13*(4), 129–134. DOI: 10.3184/095422901782775390

Davies, J. T. (1957). A quantitative kinetic theory of emulsion type, I. Physical chemistry of the emulsifying agent. *Gas/Liquid and Liquid/Liquid Interface, 1*, 426–438. Available at: https://citeseerx.ist.psu.edu/document?repid=rep1&type=pdf&doi=3247eead9a12ee37c1216fd7ab95ebd863823c88 [Accessed 21 Aug. 2023]

ECHA. (2023). *Substances restricted under REACH - Annex XVII*. [online] Available at: https://echa.europa.eu/substances-restricted-under-reach [Accessed 24 Apr. 2024]

ECRI. (2020, September 22). *Up to 70% of Chinese KN95 masks tested by ECRI don't meet minimum standards*. ECRI. [online] Available at: https://home.ecri.org/blogs/ecri-news/up-to-70-of-chinese-kn95-masks-tested-by-ecri-dont-meet-minimum-standards [Accessed 2 Feb. 2024]

Fife, G. R. (Ed.). (2021). *Greener Solvents in Conservation: An Introductory Guide*. Archetype. Available at: https://siconserve.wpengine.com/wp-content/uploads/2021/12/Greener-Solvents-in-Conservation-ONLINE-VERSION-v5.pdf

Fife, G. R., & Gill-Brown, I. (2021). Introduction. In G. R. Fife (Ed.), *Greener Solvents in Conservation: An Introductory Guide*. Archetype. Available at: https://siconserve.wpengine.com/wp-content/uploads/2021/12/Greener-Solvents-in-Conservation-ONLINE-VERSION-v5.pdf

Gold, V. (Ed.). (2019). *The IUPAC Compendium of Chemical Terminology: The Gold Book* (4th ed.). International Union of Pure and Applied Chemistry (IUPAC). DOI: 10.1351/goldbook

Great Britain: Health and Safety Executive (HSE). (2022). *Personal Protective Equipment at Work: Regulations 1992 (as Amended)* (4th ed.). HSE BOOKS. Available at: https://books.hse.gov.uk/gempdf/L25.pdf

Griffin, W. C. (1949). Classification of surface-active agents by "HLB." *Journal of the Society of Cosmetic Chemists, 1*(5), 311–326.

Haller, J., Hage, E., Ludolph, N., & Taube, A. (2020). *Minimum sample weights according to USP <41>, OIML R 76 and EURAMET cg-18.* [White paper] Available at: https://www.sartorius.com/resource/blob/580758/6e89cff42ec6efa2863140b467dc5463/minimum-weight-white-paper-en-sartorius-pdf-data.pdf [Accessed 11 Oct. 2023]

Haynes, W. M., Lide, D. R., & Bruno, T. J. (Eds.). (2017). *CRC Handbook of Chemistry and Physics: A Ready-Reference Book of Chemical and Physical Data* (2016th–2017th, 97th ed. eds.). CRC Press.

Hicks, M., & Harris, G. (2022). *Specifications and Tolerances for Reference Standard and Field Standard Weights and Measures* (No. NIST Handbook 105-2). National Institute of Standards and Technology. DOI: 10.6028/NIST.HB.105-2e2021

Ihlenfeldt, W. D., Bolton, E. E., & Bryant, S. H. (2009). The PubChem chemical structure sketcher. *Journal of Cheminformatics, 1*(1), 20. DOI: 10.1186/1758-2946-1-20

ISO. (2024, May 27). *ISO - International Organization for Standardization.* ISO. [online] Available at: https://www.iso.org/home.html [Accessed 2 Feb. 2024]

Jacobs, A. (2021, November 30). Counterfeit covid masks are still sold everywhere, despite misleading claims. *The New York Times.* [online] Available at: https://www.nytimes.com/2021/11/30/health/covid-masks-counterfeit-fake.html [Accessed 15 Jan. 2024]

Jordan, A., Stoy, P., & Sneddon, H. F. (2021). Chlorinated solvents: Their advantages, disadvantages, and alternatives in organic and medicinal chemistry. *Chemical Reviews, 121*(3), 1582–1622. DOI: 10.1021/acs.chemrev.0c00709

Klingner, T. D., & Boeniger, M. F. (2002). A critique of assumptions about selecting chemical-resistant gloves: A case for workplace evaluation of glove efficacy. *Applied Occupational and Environmental Hygiene, 17*(5), 360–367. DOI: 10.1080/10473220252864969

Landström, C., Hauxwell-Baldwin, R., Lorenzoni, I., & Rogers-Hayden, T. (2015). The (Mis)understanding of scientific uncertainty? how experts view policy-makers, the media and public. *Science as Culture, 24*(3), 276–298. DOI: 10.1080/09505431.2014.992333

Lange, N. A., & Dean, J. A. (1998). *Lange's Handbook of Chemistry* (15th ed). McGraw-Hill.

Metheny, N. A., Gunn, E. M., Rubbelke, C. S., Quillen, T. F., Ezekiel, U. R., & Meert, K. L. (2017a). Effect of pH test-strip characteristics on accuracy of readings. *Critical Care Nurse, 37*(3), 50–58. DOI: 10.4037/ccn2017199

Mizumura, M., Kubo, T., & Moriki, T. (2015a). Japanese paper: History, development and use in Western paper conservation. *Proceedings of the International Conference of the Icon Book & Paper Group*, 43–59. Available at: https://www.icon.org.uk/static/90d39d99-0aa5-426e-b5c4c1d53dd9793c/MizimuraHistory-of-Japanese-paper.pdf

Nakamura, D., Hirano, M., & Ohta, R. (2017a). Nontoxic organic solvents identified using an: A priori approach with Hansen solubility parameters. *Chemical Communications, 53*(29), 4096–4099. DOI: 10.1039/c7cc01434a

NIST. (2023). *National Institute for Standards and Technology (NIST) Chemistry WebBook.* [online] Available at: https://webbook.nist.gov/cgi/cbook.cgi?Name=EDTA&Units=SI

Occupational Safety and Health Administration (OSHA). (2023). *Personal Protective Equipment* (No. OSHA 3151-02R 2023). U.S. Department of Labor. [online] Available at: https://www.osha.gov/sites/default/files/publications/osha3151.pdf [Accessed 05 Dec. 2023]

O'Driscoll, A. (2018, December 11). *A Guide to Balance and Scale Calibration.* [online] Available at: https://labbalances.net/blogs/blog/balance-and-scale-calibration-guide

OECD, & European Union Intellectual Property Office. (2022). *Dangerous Fakes: Trade in Counterfeit Goods that Pose Health, Safety and Environmental Risks.* OECD. DOI: 10.1787/117e352b-en

Omni International Corp. (2021, June 16). *Millimeter to mil to gram: Change in the glove industry.* [online] Available at: https://omnigloves.com/2021/06/16/mil-to-gram/ [Accessed 03 Mar. 2024]

Pereira, C. S. M., Silva, V. M. T. M., & Rodrigues, A. E. (2011). Ethyl lactate as a solvent: Properties, applications and production processes – a review. *Green Chemistry, 13*(10), 2658–2671. DOI: 10.1039/c1gc15523g

Phalen, R. N. (2018). *Not All Gloves are Created Equal.* [online] Available at: https://synergist.aiha.org/201812-not-all-gloves-are-created-equal [Accessed 03 Mar. 2024]

Phalen, R. N., Le, T., & Wong, W. K. (2014). Changes in chemical permeation of disposable latex, nitrile, and vinyl gloves exposed to simulated movement. *Journal of Occupational and Environmental Hygiene, 11*(11), 716–721. DOI: 10.1080/15459624.2014.908259

Phalen, R. N., & Wong, W. K. (2015). Polymer properties associated with chemical permeation performance of disposable nitrile rubber gloves. *Journal of Applied Polymer Science, 132*(6), app.41449. DOI: 10.1002/app.41449

Phenix, A. (2007, May 2). Generic hydrocarbon solvents: A guide to nomenclature. *WAAC Newsletter, 29*(2), 13–22. Available at: https://cool.culturalheritage.org/waac/wn/wn29/wn29-2/wn29-204.pdf

Pinkert, D., & Tate, C. (2023, November 27). Stringency of us forced labor law is crucial to its success. *Sourcing Journal.* Available at: https://sourcingjournal.com/topics/thought-leadership/us-forced-labor-law-china-uflpa-customs-cbp-uyghur-genocide-469337/

Prat, D., Pardigon, O., Flemming, H.-W., Letestu, S., Ducandas, V., Isnard, P., Guntrum, E., Senac, T., Ruisseau, S., Cruciani, P., & Hosek, P. (2013a). Sanofi's solvent selection guide: A step toward more sustainable processes. *Organic Process Research & Development, 17*(12), 1517–1525. DOI: 10.1021/op4002565

Prat, D., Wells, A., Hayler, J., Sneddon, H., McElroy, C. R., Abou-Shehada, S., & Dunn, P. J. (2016a). CHEM21 selection guide of classical- and less classical-solvents. *Green Chemistry, 18*(1), 288–296. DOI: 10.1039/C5GC01008J

Precisa. (2024). *Accurate Weighing | Selecting the Right Balance | Top Tips.* [online] Available at: https://www.precisa.com/blog/toptips-to-ensure-accurate-weighing/ [Accessed 14 Jan. 2024]

PubChem. (n.d.). *PubChem Sketcher V2.4.* PubChem Sketcher. [online] Available at: https://pubchem.ncbi.nlm.nih.gov//edit3/index.html [Accessed 8 Jun. 2024]

*Standard Specification for Laboratory Glass Graduated Cylinders* (No. ASTM E1272-02(2019)). (2019). DOI: 10.1520/E1272-02R19

Stavroudis, C. (2012a). Pemulen® revised: Phuck the pH Meter. *WAAC Newsletter, 34*(2), 19. June 8, 2024. Available at: https://cool.culturalheritage.org/waac/wn/wn34/wn34-2/wn34-206.pdf

Townsend, J. (2024). *An Introduction to Materials and Chemistry* (3rd edition). Routledge, Taylor & Francis Group.

UK Department for Business and Trade. (2020, December 31). *Using the UKCA marking.* [online] Available at: https://www.gov.uk/guidance/using-the-ukca-marking [Accessed 22 Feb. 2024]

United Kingdom Accreditation Service (UKAS). (2023). *Traceability: Volumetric apparatus* (No. LAB 15; Version 5). [online] Available at: https://www.ukas.com/wp-content/uploads/2023/05/LAB-15-Traceability-Volumetric-apparatus.pdf [Accessed 22 Feb. 2024]

University of Pennsylvania's Office of Environmental Health and Radiation Safety (EHRS). (2023, May 15). *Fact Sheet: Disposable Nitrile Gloves in Chemical Labs.* [online] Available at: https://ehrs.upenn.edu/health-safety/lab-safety/chemical-hygiene-plan/fact-sheets/fact-sheet-disposable-nitrile-gloves [Accessed 05 Feb. 2024]

Westmoreland, M. B., & Breton, T. (2022, June 15). *Amazon's three-pillar strategy to fight counterfeits* [Interview]. [online] Available at: https://www.aboutamazon.eu/news/policy/amazons-three-pillar-strategy-to-fight-counterfeits

Wynne, J., Watson, K., Yesinowski, J., Young, C., Clayton, C., Nesteruk, N., Kelley, J., & Braswell, T. (2011). Report on scientific basis for paint stripping: Mechanism of methylene chloride based paint removers (No. NRL/MR/6120--11-9368; p. 73). *United States Naval Research Laboratory*. Available at: https://apps.dtic.mil/sti/tr/pdf/ADA552538.pdf

## 2 – Adhesives and Consolidants

AIC & FAIC. (2018, November 19). *Making Wheat Starch Paste.* [online video] Available at: https://www.youtube.com/watch?v=cm9yd35K4yQ [Accessed 18 Aug. 2023]

Alderson, S., Down, J. L., Maines, C. A., Williams, R. S., & Young, G. S. (2019). Potential substitutes for discontinued poly(vinyl acetate) resins used in conservation. *Journal of the American Institute for Conservation*, *58*(3), 158–179. DOI: 10.1080/01971360.2019.1589940

Arslanoglu, J. (2004). Aquazol as used in conservation practice. *WAAC Newsletter*, *26*(1). Available at: https://cool.culturalheritage.org/waac/wn/wn26/wn26-1/wn26-105.pdf

Baker, C. A. (1984). Ethylcellulose and sodium carboxymethylcellulose: An evaluation for use in paper conservation through accelerated aging. *Studies in Conservation 29*(1984): 55–59.

Belard, R., Higuchi, H., & Perry, J. (2009). Furunori (aged wheat starch paste): Challenges of production in non-traditional settings. *Journal of the Institute of Conservation*, *32*(1), 31–51. DOI: 10.1080/19455220802630735

Borges, I. da S., Casimiro, M. H., Macedo, M. F., & Sequeira, S. O. (2018). Adhesives used in paper conservation: Chemical stability and fungal bioreceptivity. *Journal of Cultural Heritage*, *34*, 53–60. DOI: 10.1016/j.culher.2018.03.027

*BPG Adhesive Recipes and Tips*. (2023, February 14). [online] Available at: https://www.conservation-wiki.com/wiki/BPG_Adhesive_Recipes_and_Tips [Accessed 05 Jun. 2023]

Bridarolli, A., Freeman, A. A., Fujisawa, N., & Łukomski, M. (2022). Mechanical properties of mammalian and fish glues over range of temperature and humidity. *Journal of Cultural Heritage*, *53*, 226–235. DOI: 10.1016/j.culher.2021.12.005

Catenazzi, K. (2017). Evaluation of the use of funori for consolidation of powdering paint layers in wall paintings. *Studies in Conservation*, *62*(2), 96–103. DOI: 10.1080/00393630.2015.1131043

Cheney, J. E. (2022). Extracting isinglass from fresh air bladders of sustainable Canadian-sourced wild Atlantic sturgeon. *Journal of the Canadian Association for Conservation*, *46*. Available at: https://www.cac-accr.ca/publications/extracting-isinglass-from-fresh-air-bladders-of-sustainable-canadian-sourced-wild-atlantic-sturgeon/

Chiantore, O., & Lazzari, M. (1996). Characterization of acrylic resins. *International Journal of Polymer Analysis and Characterization*, *2*(4), 395–408. DOI: 10.1080/10236669608033358

Davison, S. (1998). Reversible fills for transparent and translucent materials. *Journal of the American Institute for Conservation*, *37*(1), 35–47. DOI: 10.1179/019713698806082994

Down, J. L. (1984). The yellowing of epoxy resin adhesives: Report on natural dark aging. *Studies in Conservation*, *29*(2), 63–76. DOI: 10.1179/sic.1984.29.2.63

Down, J. L., Macdonald, M. A., Tétreault, J., & Williams, R. S. (1996). Adhesive testing at the Canadian Conservation Institute: An evaluation of selected poly(vinyl acetate) and acrylic adhesives. *Studies in Conservation*, *41*(1), 19–44. DOI: 10.1179/sic.1996.41.1.19

Ebert, B., Singer, B., & Grimaldi, N. (2012). Aquazol as a consolidant for matte paint on Vietnamese paintings. *Journal of the Institute of Conservation*, *35*(1), 62–76. DOI: 10.1080/19455224.2012.672813

Ebnesajjad, S. (Ed.). (2009). Chapter 5 - characteristics of adhesive materials. In *Adhesives Technology Handbook* (2nd Edition, pp. 63–135). William Andrew Publishing. DOI: 10.1016/B978-0-8155-1533-3.50008-6

Edwards, W. P. (2020, March 20). *Choosing and mixing hot hide glue*. [online] Available at: https://www.finewoodworking.com/2020/03/20/choosing-and-mixing-hot-hide-glue [Accessed 18 Aug. 2023]

Feller, R. L., & Wilt, M. (1990a). *Evaluation of Cellulose Ethers for Conservation*. The Getty Conservation Institute.

Ford, F. (1999). *I'll stick with hide glue…* [online] Available at: http://www.frets.com/FretsPages/Luthier/Data/Materials/hideglue.html [Accessed 18 Aug. 2023]

Gattuso, C., Campanella, L., Giosafatto, C. V. L., Mariniello, L., & Roviello, V. (2021). The consolidating and adhesive properties of funori: Microscopy findings on common and ancient paper samples. *Journal of Cultural Heritage*, *48*, 153–160. DOI:10.1016/j.culher.2020.11.013

Gayathri, P. (1992). Conservation of miniature paintings on paper. *Conservation of Cultural Property in India*, *25*.

Geiger, T., & Michel, F. (2005). Studies on the polysaccharide JunFunori used to consolidate matt paint. *Studies in Conservation*, *50*(3), 193–204. DOI: 10.1179/sic.2005.50.3.193

Gottsman, S. (1999). Investigation and conservation of an Egyptian mummy board from the Royal Albert Memorial Museum, Exeter. *SSCR Journal: The Quarterly News Magazine of the Scottish Society for Conservation and Restoration*, *10*.

Harrold, J., & Wyszomirska-Noga, Z. (2015). Funori: The use of a traditional Japanese adhesive in the preservation and conservation treatment of Western objects. *Adapt & Evolve 2015: East Asian Materials and Techniques in Western Conservation*, 69–79. Available at: https://www.icon.org.uk/static/1b2d61c7-53b0-4975-862a55e1ab51e619/HarroldFunori-Japanese-adhesive.pdf

Hayakawa, N., Kida, K., Ohmura, T., Yamamoto, N., Kusunoki, K., & Kawanobe, W. (2014). Characterisation of funori as a conservation material: Influence of seaweed species and extraction temperature. *Studies in Conservation*, *59*, S230–S231. DOI: 10.1179/204705814X13975704320035

Hayakawa, N., Kigawa, R., Nishimoto, T., Sakamoto, K., Fukuda, S., Kimishima, T., Oka, Y., & Kawanobe, W. (2007). Characterization of furunori (aged paste) and preparation of a polysaccharide similar to furunori. *Studies in Conservation*, *52*(3), 221–232. https://www.jstor.org/stable/20619504

Hayakawa, N., Omura, T., Hara, Y., Kusunoki, K., Kida, K., & Honda, T. (2015, June 28). Effect of extraction conditions on the physical properties of funori extracts: Temperature, seed, and water hardness. *フノリ抽出物の物性に及ぼす抽出条件の影響—温度・種・水の硬度—(早川典子、大村卓也、原由宇稀、楠京子、貴田啓子、本多貴之) 文化財保存修復学会第37回大会 京都工芸繊維大学*. 37th Conference of the Society for Conservation and Restoration of Cultural Properties, Kyoto, Japan.

Hems, E. S., & Curtis, A. E. (2015). Decolourising isinglass derived from aqua-farmed sturgeon by sonication. *Journal of the Institute of Conservation, 38*(2), 188–199. DOI: 10.1080/19455224.2015.1068199

Horie, C. V. (1987). *Materials for Conservation: Organic Consolidants, Adhesives, and Coatings*. Butterworths.

Hubbard, J. (1977). Animal glues. In I. Skeist, *Handbook of Adhesives*. New York: Van Nostrand Reinhold Co.

Karayannidou, E. G., Achilias, D. S., & Sideridou, I. D. (2006). Cure kinetics of epoxy–amine resins used in the restoration of works of art from glass or ceramic. *European Polymer Journal, 42*(12), 3311–3323. DOI: 10.1016/j.eurpolymj.2006.08.025

Kerschner, R. L., & Ravenel, N. (2006). Here we go ’round again: Cleaning linseed oil from carousel animals at the Shelburne Museum. *Journal of the American Institute for Conservation, 45*(3), 201–210. Available at: https://www.jstor.org/stable/40026691

Koob, S. P. (1979). The removal of aged shellac adhesive from ceramics. *Studies in Conservation, 24*(3), 134–135. DOI: 10.1179/sic.1979.015

Koob, S. P. (2004). Tips and tricks with epoxy and other casting and molding materials. *Objects Specialty Group Postprints, 10*, 158–172. Available at: https://resources.culturalheritage.org/osg-postprints/wp-content/uploads/sites/8/2015/02/osg010-14.pdf

Koob, S. P. (2018). Manipulating materials: Preparing and using Paraloid B-72 adhesive mixtures. *Objects Specialty Group Postprints, 25*, 1–8. Available at: https://resources.culturalheritage.org/osg-postprints/v25/koob/

Kühn, H. (1986). *Conservation and Restoration of Works of Art and Antiquities*. Butterworths.

Kwan, K. (2020). A comparison of fluids for animal glue removal from book spines. *The Book & Paper Group Annual, 39*.

Leach, A. A. (1967). Collagen chemistry in relation to isinglass and isinglass finings—a review. *Journal of the Institute of Brewing, 73*(1), 8–16. DOI: 10.1002/j.2050-0416.1967.tb03011.x

Ledoux, N., Rayner, G., Costello, S., & Chang, A. (2023). Preventive conservation, treatment, and technical study of plasticized poly(vinyl chloride) Multiples by Joseph Beuys. *Studies in Conservation, 68*(3), 343–356. DOI: 10.1080/00393630.2022.2033520

Maitland, C. (2010). Microscopy for paper conservation: Comparing various adhesives and examining wheat starch paste preparation methods. *The Book and Paper Group Annual, 29*, 129–138.

March, T., Bruno, L., Kariya, H., Ng, W., Pearlstein, E., & Stockman, H. (2006). Conservation of Assyrian reliefs at the Brooklyn Museum. *AIC Object Specialty Group Postprints*, 13, 143–167. Available at: https://faic.wpenginepowered.com/osg-postprints/wp-content/uploads/sites/8/2015/02/osg013-09.pdf

Masuda, K. (1984). Vegetable adhesives used in the workshop of the *Hyōgushi*, restorer and mounter of Japanese paintings. *Studies in Conservation, 29*(supl), 127–128. DOI: 10.1179/sic.1984.29.Supplement-1.127

Matsumaru, M. (2016, October 20). Paste-making tips with two recipes. *The Book & Paper Gathering*. Available at: https://thebookandpapergathering.org/2016/10/20/paste-making-tips-with-two-recipes/

Matsumaru, M. (2021). Wheat starch paste: A study of cooking profiles and adhesive properties across preparation recipes. *Journal of the Institute of Conservation, 44*(1), 25–46. DOI: 10.1080/19455224.2020.1864747

Matsumaru, M. (2022, January 20). Japanese horsehair sieves: Who makes them and how to use them? *The Book & Paper Gathering*. Available at: https://thebookandpapergathering.org/2022/01/20/japanese-horsehair-sieves-who-makes-them-and-how-to-use-them/

Melià-Angulo, A., Fuster-López, L., & Escuder, A. V. (2017). Study of the mechanical properties of selected animal glues and their implication when designing conservation strategies. *CeROArt. Conservation, exposition, Restauration d'Objets d'Art, HS.* DOI: 10.4000/ceroart.5152

Miller, B. F., & Root, W. (1991). Long-term storage of wheat starch paste. *Studies in Conservation, 36*(2), 85–92. DOI: 10.2307/1506332

Mizumura, M., Kubo, T., & Moriki, T. (2015b). Japanese paper: History, development and use in Western paper conservation. *Proceedings of the International Conference of the Icon Book & Paper Group*, 43–59. Available at: https://www.icon.org.uk/static/90d39d99-0aa5-426e-b5c4c1d53dd9793c/MizimuraHistory-of-Japanese-paper.pdf

Morita, T. (1984). 'Nikawa' – traditional production of animal glue in Japan. *Studies in Conservation, 29*, 121–122. DOI: 10.1179/sic.1984.29.Supplement-1.121

Mucci, P. (1975). Technical notes: Paste recipes. *The Mid-Atlantic Archivist, 4*(3), 7–8.

Nogueira, G. F., Fakhouri, F. M., & De Oliveira, R. A. (2018). Extraction and characterization of arrowroot (Maranta arundinaceae L.) starch and its application in edible films. *Carbohydrate Polymers, 186*, 64–72. DOI: 10.1016/j.carbpol.2018.01.024

Ntasi, G., Sbriglia, S., Pitocchi, R., Vinciguerra, R., Melchiorre, C., Dello Ioio, L., Fatigati, G., Crisci, E., Bonaduce, I., Carpentieri, A., Marino, G., & Birolo, L. (2022a). Proteomic characterization of collagen-based animal glues for restoration. *Journal of Proteome Research.* DOI: 10.1021/acs.jproteome.2c00232

Petukhova, T., & Bonadies, S. D. (1993). Sturgeon glue for painting consolidation in Russia. *Journal of the American Institute for Conservation, 32*(1), 23–31. DOI: 10.1179/019713693806066483

Poulis, J. A., Mosleh, Y., Cansell, E., Cimino, D., Ploeger, R., René de la Rie, E., McGlinchey, C. W., & Seymour, K. (2022). Mechanical and physical characterization of natural and synthetic consolidants. *International Journal of Adhesion and Adhesives, 117*, 103015. DOI: 10.1016/j.ijadhadh.2021.103015

Prat, D., Pardigon, O., Flemming, H.-W., Letestu, S., Ducandas, V., Isnard, P., Guntrum, E., Senac, T., Ruisseau, S., Cruciani, P., & Hosek, P. (2013b). Sanofi's solvent selection guide: A step toward more sustainable processes. *Organic Process Research & Development, 17*(12), 1517–1525. DOI: 10.1021/op4002565

Prat, D., Wells, A., Hayler, J., Sneddon, H., McElroy, C. R., Abou-Shehada, S., & Dunn, P. J. (2016b). CHEM21 selection guide of classical- and less classical-solvents. *Green Chemistry, 18*(1), 288–296. DOI: 10.1039/C5GC01008J

Reinartz, R., & Slavcheva, P. (2016). *Saving Sturgeons: A global report on their status and suggested conservation strategy.* [online] Available at: https://wwfint.awsassets.panda.org/downloads/Saving_sturgeons_A_global_report_on_their_status_and_suggested_conservation_strategy.pdf

Rivers, S., & Umney, N. (2013a). *Conservation of Furniture.* Routledge.

Schellmann, N. C. (2007). Animal glues: A review of their key properties relevant to conservation. *Studies in Conservation, 52*(supl), 55–66. DOI: 10.1179/sic.2007.52.supplement-1.55

Schilling, M. R. (1989). The glass transition of materials used in conservation. *Studies in Conservation, 34*(3), 110–116. DOI: 10.1179/sic.1989.34.3.110

Schuck, T., Fischer, A., & Schwahn, B. (2019). Acrylic barrier surface coatings for epoxy resin adhesive bonds in glass conservation: Evaluation of bond strength and reversibility. *Recent Advances in Glass and Ceramics Conservation 2019: Interim Meeting of the ICOM-CC Glass and Ceramics Working Group and Icon Ceramics and Glass Group Conference.*

Sheesley, S. (2011). Practical applications of Lascaux acrylic dispersions in paper conservation. *The Book and Paper Group Annual, 30*, 79–81.

Shockey, H. (2017). Structural repair of plaster sculpture using polyvinyl butyral resin. *Objects Specialty Group Postprints, 24*, 136–146.

Shulman, K., & Matsen, C. (2022). Agar bioplastic: Understanding its degradation and exploring its viability as a fill for glass. *Student Papers & Posters*. ANAGPIC, the Association of North American Graduate Programs in Conservation, Available at: https://faic.wpenginepowered.com/anagpic-student-papers/wp-content/uploads/sites/11/2023/08/Shulman_Agar-Bioplastic_2022-1.pdf

Sideridou, I. D., Vouvoudi, E. C., & Papadopoulos, G. D. (2016). Epoxy polymer Hxtal NYL-1™ used in restoration and conservation: Irradiation with short and long wavelengths and study of photo-oxidation by FT-IR spectroscopy. *Journal of Cultural Heritage*, *18*, 279–289. DOI: 10.1016/j.culher.2015.09.005

Skeist, I. (1977). *Handbook of Adhesives*. New York: Van Nostrand Reinhold Co.

Spicer, G. (1992). *Sticky Business, the Removal of Hide Glue from Silk*. DOI: 10.13140/RG.2.2.36749.23525

Spirydowicz, K. E., Simpson, E., Blanchette, R. A., Schniewind, A. P., Toutloff, M. K., & Murray, A. (2001). Alvar and Butvar: The use of polyvinyl acetal resins for the treatment of the wooden artifacts from Gordion, Turkey. *Journal of the American Institute for Conservation*, *40*(1), 43–57. DOI: 10.2307/3180012

Stavroudis, C., & Blank, S. (1989a, May). Solvents & Sensibility. *WAAC Newsletter*, *11*(1), 2–10. https://cool.culturalheritage.org/waac/wn/wn11/wn11-2/wn11-202.html

Stratton, L. (2021). Aquazol stabilized hydrogen peroxide: An innovative combination. *Polymer Chemistry Innovations Inc.*

Swider, J. R., & Smith, M. (2005). Funori: Overview of a 300-year-old consolidant. *Journal of the American Institute for Conservation*, *44*(2), 117–126. DOI: 10.1179/019713605806082329

Tuvikene, R., Robal, M., Fujita, D., Saluri, K., Truus, K., Tashiro, Y., Ogawa, H., & Matsukawa, S. (2015). Funorans from Gloiopeltis species. Part I. Extraction and structural characteristics. *Food Hydrocolloids*, *43*, 481–492. DOI: 10.1016/j.foodhyd.2014.07.010

van Dyke, Y. (2009a). Sacred leaves: The conservation and exhibition of early Buddhist manuscripts on palm leaves. *The Book and Paper Group Annual*, *28*. https://cool.culturalheritage.org/coolaic/sg/bpg/annual/v28/bpga28-17.pdf

Vinçotte, A., Beauvoit, E., Boyard, N., & Guilminot, E. (2019). Effect of solvent on Paraloid® B72 and B44 acrylic resins used as adhesives in conservation. *Heritage Science*, *7*(1). DOI: 10.1186/s40494-019-0283-9

Wolfe, J., & Grayburn, R. (2017). A review of the development and testing of Incralac lacquer. *Journal of the American Institute for Conservation*, *56*(3–4), 225–244. DOI: 10.1080/01971360.2017.1362863

Zeng, M., Morris, C. F., Batey, I. L., & Wrigley, C. W. (1997a). Sources of variation for starch gelatinization, pasting, and gelation properties in wheat. *Cereal Chemistry*, *74*(1), 63–71. DOI: 10.1094/CCHEM.1997.74.1.63

## 3 – Clays and Gels

Acebal, S., & Vico, L. (2017). Acid-base properties of aqueous suspensions of homoionic sepiolite and palygorskite. *Natural Resources*, *8*(6), 432–444. DOI: 10.4236/NR.2017.86028

Aguilar Sánchez, B., & Buentello Martínez, R. (2017). Agarose gels with methyl ethyl ketone for cleaning a 19th-century document. In L. V. Angelova, B. Ormsby, J. Townsend, & R. Wolbers (Eds.), *Gels in the Conservation of Art*. (pp. 107–109). Archetype.

Akkineni, A. R., Elci, B. S., Lode, A., & Gelinsky, M. (2022). Addition of high acyl gellan gum to low acyl gellan gum enables the blends 3D bioprintable. *Gels*, *8*(4), 199. DOI: 10.3390/gels8040199

Al-Emam, E., Motawea, A. G., Janssens, K., & Caen, J. (2019). Evaluation of polyvinyl alcohol–borax/agarose (PVA–B/AG) blend hydrogels for removal of deteriorated consolidants from ancient Egyptian wall paintings. *Heritage Science*, *7*(1), 22. DOI: 10.1186/s40494-019-0264-z

Al-Emam, E., Soenen, H., Caen, J., & Janssens, K. (2020). Characterization of polyvinyl alcohol-borax/agarose (PVA-B/AG) double network hydrogel utilized for the cleaning of works of art. *Heritage Science*, *8*(1). DOI: 10.1186/s40494-020-00447-3

Angelova, L. V., Berrie, B. H., De Ghetaldi, K., Kerr, A., & Weiss, R. G. (2015). Partially hydrolyzed poly(vinyl acetate)-borax-based gel-like materials for conservation of art: Characterization and applications. *Studies in Conservation*, *60*(4), 227–244. DOI: 10.1179/2047058413Y.0000000112

Angelova, L. V., Berrie, B., & Weiss, R. (2017a). Poly (vinyl alcohol)-borax 'gels': A flexible cleaning option. In L. V. Angelova, B. Ormsby, J. Townsend, & R. Wolbers (Eds.), *Gels in the Conservation of Art*. (pp. 231–236). Archetype.

Angelova, L. V., Ormsby, B., Townsend, J., & Wolbers, R. (Eds.). (2017b). *Gels in the Conservation of Art*. Archetype.

Angelova, L. V., Terech, P., Natali, I., Dei, L., Carretti, E., & Weiss, R. G. (2011). Cosolvent gel-like materials from partially hydrolyzed poly(vinyl acetate)s and borax. *Langmuir*, *27*(18), 11671–11682. DOI: 10.1021/la202179e

Arab, M., Yousefi, M., Khanniri, E., Azari, M., Ghasemzadeh-Mohammadi, V., & Mollakhalili-Meybodi, N. (2023). A comprehensive review on yogurt syneresis: Effect of processing conditions and added additives. *Journal of Food Science and Technology*, *60*(6), 1656–1665. DOI: 10.1007/s13197-022-05403-6

Aruldass, S., Mathivanan, V., Mohamed, A. R., & Tye, C. T. (2019). Factors affecting hydrolysis of polyvinyl acetate to polyvinyl alcohol. *Journal of Environmental Chemical Engineering*, *7*(5), 103238. DOI: 10.1016/j.jece.2019.103238

Baglioni, P., Berti, D., Bonini, M., Carretti, E., Perez, M. D. C. C., Chelazzi, D., Dei, L., Fratini, E., Giorgi, R., Natali, I., & Arroyo, M. C. (2012). Gels for the conservation of cultural heritage. *MRS Online Proceedings Library (OPL)*, *1418*, mrsf11-1418-ll02-01. DOI: 10.1557/OPL.2012.97

Baglioni, P., & Chelazzi, D. (Eds.). (2013). *Nanoscience for the Conservation of Works of Art*. RSC Publishing.

Barberà Giné, A., & Marín Ortega, S. (2017). The removal of Paraloid-B72 coatings with aqueous gelled systems: Roman frescoes from Empúries, Catalonia. In L. V. Angelova, B. Ormsby, J. Townsend, & R. Wolbers (Eds.), *Gels in the Conservation of Art*. (pp. 363–366). Archetype.

Barbisan, S. (2018). Tip: Local cleaning with gels: Acknowledging the challenges and successes. *AIC Book and Paper Group Annual*, *37*, 152–155. Available at: https://cool.culturalheritage.org/coolaic/sg/bpg/annual/v37/bpga37-24.pdf

Barbisan, S., & Dupont, A.-L. (2017). Local cleaning of tidelines on paper using rigid gels: The influence of pH and conductivity. In L. V. Angelova, B. Ormsby, J. Townsend, & R. Wolbers (Eds.), *Gels in the Conservation of Art*. (pp. 113–115). Archetype.

Barkovic, M., Diamond, O., & Cross, M. (2017). The use of agar gel for treating water stains on an acrylic canvas. In L. V. Angelova, B. Ormsby, J. Townsend, & R. Wolbers (Eds.), *Gels in the Conservation of Art*. (pp. 51–56). Archetype.

Beaumont, M., Tran, R., Vera, G., Niedrist, D., Rousset, A., Pierre, R., Shastri, V. P., & Forget, A. (2021). Hydrogel-forming algae polysaccharides: From seaweed to biomedical applications. *Biomacromolecules*, *22*(3), 1027–1052. DOI: 10.1021/acs.biomac.0c01406

Berlangieri, C., Andrina, E., Matarrese, C., Carretti, E., Traversi, R., Severi, M., Chelazzi, D., Dei, L., & Baglioni, P. (2017a). Chelators confined into 80pvac-borax highly viscous dispersions for the removal of gypsum degradation layers. *Pure and Applied Chemistry*, *89*(1), 97–109. DOI: 10.1515/pac-2016-0815

Bertasa, M., Canevali, C., Sansonetti, A., Lazzari, M., Malandrino, M., Simonutti, R., & Scalarone, D. (2021). An in-depth study on the agar gel effectiveness for built heritage cleaning. *Journal of Cultural Heritage*, *47*, 12–20. DOI: 10.1016/j.culher.2020.10.007

Bertasa, M., Chiantore, O., Poli, T., Riedo, C., Di Tullio, V., Canevali, C., Sansonetti, A., & Scalarone, D. (2017). A study of commercial agar gels as cleaning materials. In L. V. Angelova, B. Ormsby, J. Townsend, & R. Wolbers (Eds.), *Gels in the Conservation of Art*. (pp. 11–18). Archetype.

Bertasa, M., Dodero, A., Alloisio, M., Vicini, S., Riedo, C., Sansonetti, A., Scalarone, D., & Castellano, M. (2020a). Agar gel strength: A correlation study between chemical composition and rheological properties. *European Polymer Journal, 123*. DOI: 10.1016/j.eurpolymj.2019.109442

Bertasa, M., Poli, T., Riedo, C., Di Tullio, V., Capitani, D., Proietti, N., Canevali, C., Sansonetti, A., & Scalarone, D. (2018). A study of non-bounded/bounded water and water mobility in different agar gels. *Microchemical Journal, 139*. DOI: 10.1016/j.microc.2018.03.016

Bertasa, M., Ricci, C., Scarcella, A., Zenucchini, F., Pellis, G., Croveri, P., & Scalarone, D. (2020b). Overcoming challenges in street art murals conservation: A comparative study on cleaning approach and methodology. *Coatings, 10*(11), 1–23. DOI: 10.3390/coatings10111019

Berzioli, M., Botti, L., Casoli, A., Corazza, A., Cremonesi, P., Iannuccelli, S., Isca, C., Placido, M., Residori, L., Ruggiero, D., Sotgiu, S., & Tireni, L. (2011). Evaluation of cleaning and chemical stabilization of paper treated with a rigid hydrogel of gellan gum by means of chemical and physical analyses. *Proceedings of ICOM-CC's 16th Triennial Conference, 602*, 1–11.

Blüher, A., Banik, G., Maurer, K.-H., & Thobois, E. (1996a). The application of enzyme-containing methylcellulose gels for the removal of starch-based adhesives in albums. In J. Bridgland (Ed.), *ICOM Committee for Conservation 11th Triennial Meeting Edinburgh 1-6 September 1996*. Earthscan Ltd. Available at: https://www.icom-cc-publications-online.org/2990/The-Application-of-Enzyme-containing-Methylcellulose-Gels-for-the-Removal-of-Starch-based-Adhesives-in-Albums

Bollard, C., Balcar, N., Barabant, G., Kuperholc, S. (2013). Effects of cleaning gel systems on plasticized poly(vinyl chloride): New insights. *Future Talks 013*.

Bollard, C., Kuperholc, S., Balcar, N., Barabant, G. (2011). Effects of cleaning gel systems on plastics: A preliminary study on plasticized poly(vinyl chloride). *Future Talks 011*. Technology and Conservation of Modern Materials in Design, Munich (Germany).

Bonelli, N., Poggi, G., Chelazzi, D., Giorgi, R., & Baglioni, P. (2019). Poly(vinyl alcohol)/poly(vinyl pyrrolidone) hydrogels for the cleaning of art. *Journal of Colloid and Interface Science, 536*, 339–348. DOI: 10.1016/j.jcis.2018.10.025

Boral, S., Saxena, A., & Bohidar, H. B. (2010). Syneresis in agar hydrogels. *International Journal of Biological Macromolecules, 46*(2), 232–236. DOI: 10.1016/j.ijbiomac.2009.12.008

Bourgès, A., & Vergès-Belmin, V. (2008). A new methodology to determine rheologic behavior and mechanical properties of desalination poultices. *Proceedings of the 11th International Congress on Deterioration and Conservation of Stone*, 581–588.

Brunchi, C.-E., Bercea, M., Morariu, S., & Dascalu, M. (2016). Some properties of xanthan gum in aqueous solutions: Effect of temperature and pH. *Journal of Polymer Research, 23*(7), 123. DOI: 10.1007/s10965-016-1015-4

Canevali, C., Fasoli, M., Bertasa, M., Botteon, A., Colombo, A., Di Tullio, V., Capitani, D., Proietti, N., Scalarone, D., & Sansonetti, A. (2016). A multi-analytical approach for the study of copper stain removal by agar gels. *Microchemical Journal, 129*, 249–258. DOI: 10.1016/j.microc.2016.07.007

Carnazza, P., Francone, S., Clarricoates, R., Dowding, H., & Wright, A. (Eds.) (2019). *Using High Molecular Weight Polysaccharides to Clean Vinyl Paintings: A Case Study on a Mixed-Media Contemporary Artwork and Experimental Tests*. Archetype, Institute of Conservation Paintings Group.

Carretti, E., Bonini, M., Dei, L., Berrie, B. H., Angelova, L. V., Baglioni, P., & Weiss, R. G. (2010). New frontiers in materials science for art conservation: Responsive gels and beyond. *Accounts of Chemical Research*, *43*(6), 751–760. DOI: 10.1021/ar900282h

Carretti, E., Dei, L., Macherelli, A., & Weiss, R. G. (2004). Rheoreversible polymeric organogels: The art of science for art conservation. *Langmuir*, *20*(20), 8414–8418. DOI: 10.1021/la0495175

Carretti, E., Grassi, S., Cossalter, M., Natali, I., Caminati, G., Weiss, R. G., Baglioni, P., & Dei, L. (2009). Poly (vinyl alcohol)-borate hydro/cosolvent gels: Viscoelastic properties, solubilizing power, and application to art conservation. *Langmuir*, *25*(15), 8656–8662. DOI: 10.1021/la804306w

Cassanelli, M., Norton, I., & Mills, T. (2018a). Role of gellan gum microstructure in freeze drying and rehydration mechanisms. *Food Hydrocolloids*, *75*, 51–61. DOI: 10.1016/j.foodhyd.2017.09.013

Cassanelli, M., Prosapio, V., Norton, I., & Mills, T. (2018b). Acidified/basified gellan gum gels: The role of the structure in drying/rehydration mechanisms. *Food Hydrocolloids*, *82*, 346–354. DOI: 10.1016/j.foodhyd.2018.04.024

Chen, Q., Wen, W. Y., Qiu, F. X., Xu, J. C., Yu, H. Q., Chen, M. L., & Yang, D. Y. (2016). Preparation and application of modified carboxymethyl cellulose Si/polyacrylate protective coating material for paper relics. *Chemical Papers*, *70*(7), 946–959. DOI: 10.1515/chempap-2016-0029

Crawford, S. (2024, February 12). *Carbopol polymers for gels* [Personal communication].

Cremonesi, P., & Casoli, A. (2017). Thermo-reversible rigid agar hydrogels: Their properties and action in cleaning. In L. V. Angelova, B. Ormsby, J. Townsend, & R. Wolbers (Eds.), *Gels in the Conservation of Art* (pp 19–28). Archetype.

Cremonesi, P., Mecklenburg, M., Charola, A. E., & Koestler, R. (Eds.) (2013). *Rigid Gels and Enzyme Cleaning*. Smithsonian Institution Scholarly Press.

Cushman, M. (2024, May 22). *Displayable Colored Gels?* [Reply]. Global Conservation Forum (ConsDistList). [online] Available at: https://community.culturalheritage.org/discussion/displayable-colored-gels?ReturnUrl=%2fcommunities%2fcommunity-home%2fdigestviewer%3fListKey%3d757a8f16-505f-4323-8e74-e376757aa9f7%26CommunityKey%3dea3d002c-9fc3-4446-b7d2-c308f5faed13

da Silveira, L. (1997). A Note on the Poultice Cleaning of Feathers Using Laponite RD Gel. *Studies in Conservation*, *42*(1), 11–16. DOI: 10.2307/1506571

Deckner, G. (2015, May 1). *Xanthan Gum: A Workhorse Polysaccharide*. [online] Available at: https://www.ulprospector.com/knowledge/2368/pcc-xanthan-gum-workhorse-polysaccharide/ [Accessed 12 Dec. 2023]

deGhetaldi, K., Baade, B., Voras, Z., & Wroczynski, E. (2017). Resurrecting a giant: Using solvent gels and aqueous systems to restore Villanova University's Triumph of David. In L. V. Angelova, B. Ormsby, J. Townsend, & R. Wolbers (Eds.), *Gels in the Conservation of Art*. (pp. 157–164). Archetype.

Delattre, C., Bouvet, S., & Le Bourg, É. (2017). Gellan gum and agar compared to aqueous immersion for cleaning paper. In L. V. Angelova, B. Ormsby, J. Townsend, & R. Wolbers (Eds.), *Gels in the Conservation of Art*. (pp. 57–61). Archetype.

Dhiaa, A. H. (2020). The temperature effect on the viscosity and density of xanthan gum solution. *Kufa Journal of Engineering*, *3*, 17–30. DOI: 10.30572/2018/KJE/321258

Doehne, E. F., & Price, C. A. (2010). *Stone Conservation: An Overview of Current Research* (2nd ed). Getty Conservation Institute.

Doherty, E. (2017). The removal of lead-and oil-based overpaint from a plaster cast of Hermes Fastening his Sandal. In L. V. Angelova, B. Ormsby, J. Townsend, & R. Wolbers (Eds.), *Gels in the Conservation of Art* (pp. 122–126). Archetype.

Duncan, T. T., Berrie, B. H., & Weiss, R. G. (2017). A comparison between gel and swab cleaning: Physical changes to delicate surfaces. In L. V. Angelova, B. Ormsby, J. Townsend, & R. Wolbers (Eds.), *Gels in the Conservation of Art* (pp. 250–256). Archetype.

Entwistle, R. (2017). Low budget thixotropic gels. *ICON News: The Magazine of the Institute of Conservation, 72*, 33–35.

Eriksson, H., Wedberg, I., Nessow, J., & Bronmark-Thorlund, M. (2017). The use of Nanorestore gels in the conservation of lime-based wall paintings. In L. V. Angelova, B. Ormsby, J. Townsend, & R. Wolbers (Eds.), *Gels in the Conservation of Art* (pp. 270–273). Archetype.

Faust, E. (2017). Cleaning assessment of a vinyl dispersion paint: Comparing Velvesil Plus gel, microemulsions and aqueous solutions. In L. V. Angelova, B. Ormsby, J. Townsend, & R. Wolbers (Eds.), *Gels in the Conservation of Art* (pp. 218–222). Archetype.

Feller, R. L., & Wilt, M. (1990b). *Evaluation of Cellulose Ethers for Conservation*. The Getty Conservation Institute.

Fiskovic, L. L. (2017). Carbopol gel as a carrier for oxidizing agents in discoloured lead white conversion treatment. In L. V. Angelova, B. Ormsby, J. Townsend, & R. Wolbers (Eds.), *Gels in the Conservation of Art* (pp. 181–183). Archetype.

Frost, R. L., & Kristof, J. (2004). Raman and infrared spectroscopic studies of kaolinite surfaces modified by intercalation. In F. Wypych & K. G. Satyanarayana (Eds.), *Interface Science and Technology* (Vol. 1, pp. 184–215). Elsevier. DOI: 10.1016/S1573-4285(04)80041-3

Gao, S., Guo, J., Wu, L., & Wang, S. (2008). Gelation of konjac glucomannan cross-linked by organic borate. *Carbohydrate Polymers, 73*(3), 498–505. DOI: 10.1016/j.carbpol.2007.12.013

Giordano, A., Caruso, M. R., & Lazzara, G. (2022). New tool for sustainable treatments: Agar spray—research and practice. *Heritage Science, 10*(1), 123. DOI: 10.1186/s40494-022-00756-9

Giordano, A., & Cremonesi, P. (2021). New methods of applying rigid agar gels: From tiny to large-scale surface areas. *Studies in Conservation, 66*(8). DOI: 10.1080/00393630.2020.1848272

Giorgi, R., Baglioni, M., & Baglioni, P. (2017). Nanofluids and chemical highly retentive hydrogels for controlled and selective removal of overpaintings and undesired graffiti from street art. *Analytical and Bioanalytical Chemistry, 409*(15), 3707–3712. DOI: 10.1007/s00216-017-0357-z

Giraud, T., Gomez, A., Lemoine, S., Pelé-Meziani, C., Raimon, A., & Guilminot, E. (2021). Use of gels for the cleaning of archaeological metals. Case study of silver-plated copper alloy coins. *Journal of Cultural Heritage, 52*, 73–83. DOI: 10.1016/j.culher.2021.08.014

Glenn, S., Hackett, J., Haldane, E.-A., & Im, S. (2015a). Borrowing from the neighbours: Using the technology of other disciplines to treat difficult textile conservation problems. In A. Fairhurst (Ed.), *Learning Curve: Education, Experience, Reflection. Postprints* (pp. 33–42). ICON. Available at: https://www.icon.org.uk/static/dd33714f-f5b2-4f69-8b1eb6d866d4a8b0/2015learningcurveforum.pdf

Guilminot, E. (2023). The use of hydrogels in the treatment of metal cultural heritage objects. *Gels, 9*(3), 191. DOI: 10.3390/gels9030191

Gulotta, D., Saviello, D., Gherardi, F., Toniolo, L., Anzani, M., Rabbolini, A., & Goidanich, S. (2014). Setup of a sustainable indoor cleaning methodology for the sculpted stone surfaces of the Duomo of Milan. *Heritage Science, 2*(1). DOI: 10.1186/2050-7445-2-6

Gun'ko, V. M., Savina, I. N., & Mikhalovsky, S. V. (2017). Properties of water bound in hydrogels. *Gels, 3*(4), 37. DOI: 10.3390/gels3040037

Henniges, U., Brückle, I., Khaliliyan, H., & Böhmdorfer, S. (2024). Gellan residues on paper: Quantification and implication for paper conservation. *Heritage Science, 12*(1), 54. DOI: 10.1186/s40494-024-01160-1

Henry, A., & Pearce, J. (Eds.). (2008). *Stone Conservation: Principles and Practice*. Donhead.

Huggett, J. M. (2005). Clay minerals. In R. C. Selley, L. R. M. Cocks, & I. R. Plimer (Eds.), *Encyclopedia of Geology* (pp. 358–365). Elsevier. DOI: 10.1016/B0-12-369396-9/00273-2

Iannuccelli, S., & Sotgiu, S. (2010). Wet treatments of works of art on paper with rigid gellan gels. *The Book & Paper Group Annual*, *29*.

Isaksson, C. (2017). Solvent gel versus solvent poultice: Evaluating two techniques for the removal of pressure-sensitive tape stains from paper. In L. V. Angelova, B. Ormsby, J. Townsend, & R. Wolbers (Eds.), *Gels in the Conservation of Art* (pp. 126–128). Archetype.

Jatav, S., & Joshi, Y. M. (2014). Chemical stability of Laponite in aqueous media. *Applied Clay Science*, *97–98*, 72–77. DOI: 10.1016/j.clay.2014.06.004

Jia, Y., Sciutto, G., Botteon, A., Conti, C., Focarete, M. L., Gualandi, C., Samorì, C., Prati, S., & Mazzeo, R. (2021a). Deep eutectic solvent and agar: A new green gel to remove proteinaceous-based varnishes from paintings. *Journal of Cultural Heritage*, *51*, 138–144. DOI: 10.1016/j.culher.2021.08.001

Kanth, A. P., & Pandey, S. C. (2016). Recent advances in gel-based systems for cleaning of works of art. *Conservation of Cultural Property in India*, *41*.

Kavda, S., Golfomitsou, S., & Richardson, E. (2017). The effect of gelling agents and solvents on poly(methyl methacrylate) surfaces: A comparative study. In L. V. Angelova, B. Ormsby, J. Townsend, & R. Wolbers (Eds.), *Gels in the Conservation of Art* (pp. 331–336). Archetype.

Lee, C., Di Turo, F., Vigani, B., Weththimuni, M. L., Rossi, S., Beltram, F., Pingue, P., Licchelli, M., Malagodi, M., Fiocco, G., & Volpi, F. (2022a). Biopolymer gels as a cleaning system for differently featured wooden surfaces. *Polymers*, *15*(1), 36. DOI: 10.3390/polym15010036

Lee, L.-M., Rogers, P., Oakley, V., & Navarro, J. (1997). Investigations into the use of laponite as a poulticing material in ceramics conservation. *Conservation Journal*, *22*. Available at: http://www.vam.ac.uk/content/journals/conservation-journal/issue-22/investigations-into-the-use-of-laponite-as-a-poulticing-material-in-ceramics-conservation/

Lemiski, S. (1998). An investigation of poultice materials for textile conservation. *Textile Conservation Newsletter.*, 34, 1–15.

Leroux, M. (2016). Soft matter: Gel development for conservation treatment Gellan gum and Nanorestore Gel®. *The Book & Paper Group Annual*, *35*.

Ling, D. (1991). Laponite poulticing. *Conservation News (United Kingdom Institute for Conservation of Historic and Artistic Works)*, *46*.

Lubelli, B., & van Hees, R. P. J. (2010). Desalination of masonry structures: Fine tuning of pore size distribution of poultices to substrate properties. *Journal of Cultural Heritage*, *11*(1), 10–18. DOI: 10.1016/j.culher.2009.03.005

Lubrizol. (2011a). 3 - *Polymer Handling and Storage* (Manufacturer Brochure No. Bulletin 3; Pharmaceutical Bulletin).

Lubrizol. (2011b). 4 - *Dispersion Techniques for Lubrizol Pharmaceutical Polymers* (Pharmaceutical Bulletin).

Lubrizol. (2011c). 5 - *Neutralization Procedures* (Manufacturer Brochure No. Bulletin 5; Pharmaceutical Bulletin).

Lubrizol. (2011d). 6 - *Thickening Properties* (Manufacturer Brochure No. Bulletin 6; Pharmaceutical Bulletin).

Magee, C. (2019). High acyl Gellan gum for parchment conservation. *AIC Book and Paper Group Annual*, *38*, 112–119.

Maheux, A. F. (2015). Cross-disciplinary uses for Gellan gum in conservation. *The Book and Paper Group Annual*, *34*, 69–79. Available at: https://cool.culturalheritage.org/coolaic/sg/bpg/annual/v34/bpga34-10.pdf

Mao, R., Tang, J., & Swanson, B. G. (2001). Water holding capacity and microstructure of Gellan gels. *Carbohydrate Polymers*, *46*(4), 365–371. DOI: 10.1016/S0144-8617(00)00337-4

Markevičius, T., Syversen, T., Chan, E., Olsson, N., Hilby, C. S., & Šimaitė, R. (2017a). Cold, warm, warmer: Use of precision heat transfer in the optimization of hydrolytic enzyme and hydrogel cleaning systems. In L. V. Angelova, B. Ormsby, J. Townsend, & R. Wolbers (Eds.), *Gels in the Conservation of Art* (pp. 68–72). Archetype.

Mazzuca, C., Micheli, L., Cervelli, E., Basoli, F., Cencetti, C., Coviello, T., Iannuccelli, S., Sotgiu, S., & Palleschi, A. (2014a). Cleaning of paper artworks: Development of an efficient gel-based material able to remove starch paste. *ACS Applied Materials and Interfaces*, *6*(19), 16519–16528. DOI: 10.1021/am504295n

Miller, Z., Whitby, G., & Garside, P. (2017). Investigating the ability of phytate gel systems to treat iron gall ink at the British Library. In L. V. Angelova, B. Ormsby, J. Townsend, & R. Wolbers (Eds.), *Gels in the Conservation of Art*. (pp.77–81) Archetype.

Mina, L. (2020a). Foxy underpants: Or the use of chelators and enzymes to reduce foxing stains on early nineteenth century men's linen underpants. *Journal of the American Institute for Conservation*, *59*(1), 3–17. DOI: 10.1080/01971360.2019.1674604

Moreira, H. R., da Silva, L. P., Reis, R. L., & Marques, A. P. (2020). Tailoring gellan gum spongy-like hydrogels' microstructure by controlling freezing parameters. *Polymers*, *12*(2), 329. DOI: 10.3390/polym12020329

Narayanan, J., Xiong, J.-Y., & Liu, X.-Y. (2006). Determination of agarose gel pore size: Absorbance measurements vis a vis other techniques. *Journal of Physics: Conference Series*, *28*, 83–86. DOI: 10.1088/1742-6596/28/1/017

Natali, I., Carretti, E., Angelova, L., Baglioni, P., Weiss, R. G., & Dei, L. (2011). Structural and mechanical properties of "peelable" organoaqueous dispersions with partially hydrolyzed poly(vinyl acetate)-borate networks: Applications to cleaning painted surfaces. *Langmuir*, *27*(21), 13226–13235. DOI: 10.1021/la2015786

Naunton, M. (2009). Use of gel poultices for adhesive removal. *The Textile Specialty Group Postprints (AIC)*, *19*.

Nussinovitch, A., Velez-Silvestre, R., & Peleg, M. (1993). Compressive characteristics of freeze-dried agar and alginate gel sponges. *Biotechnology Progress*, *9*(1), 101–104. DOI: 10.1021/bp00019a015

Peranteau, A., & Stevens, S. C., (Book Compiler). (2013). Gellan gum as a material for local stain reduction. In *Conserving Modernity: The Articulation of Innovation*. North American Textile Conservation Conference.

Pernodet, N., Maaloum, M., & Tinland, B. (1997). Pore size of agarose gels by atomic force microscopy. *Electrophoresis*, *18*(1), 55–58. DOI: 10.1002/elps.1150180111

Petrella, G., Mazzuca, C., Micheli, L., Cervelli, E., De Fazio, D., Iannuccelli, S., Sotgiu, S., Palleschi, G., & Palleschi, A. (2016a). A new sustainable and innovative work for paper artworks cleaning process: Gellan hydrogel combined with hydrolytic enzymes. *International Journal of Conservation Science*, *7*(special 1), 273–280.

Picone, C. S. F., & Cunha, R. L. (2011). Influence of pH on formation and properties of Gellan gels. *Carbohydrate Polymers*, *84*(1), 662–668. DOI: 10.1016/j.carbpol.2010.12.045

Prestowitz, B. (2017). An initial assessment of local stain reduction using chelate-containing Gellan gum gels in paper conservation. In L. V. Angelova, B. Ormsby, J. Townsend, & R. Wolbers (Eds.), *Gels in the Conservation of Art* (pp. 142–144). Archetype.

Prosapio, V. T., Norton, I., & Lopez-Quiroga, E. (2020). Freeze-dried gellan gum gels as vitamin delivery systems: Modelling the effect of pH on drying kinetics and vitamin release mechanisms. *Foods*, *9*(3), 329. DOI: 10.3390/foods9030329

Puoti, F., Jervis, A. V., Ciabattoni, R., Cossa, E., Di Giovanni, A., Giuliani, M. R., Guida, G., & Ioele, M. (2017). Evaluation of leather cleaning with a rigid hydrogel of gellan gum on two composite Amharic shields from the Museo Nazionale Preistorico Etnografico 'Luigi Pigorini', Rome. In L. V. Angelova, B. Ormsby, J. Townsend, & R. Wolbers (Eds.), *Gels in the Conservation of Art* (pp. 82–86). Archetype.

Qiao, D., Shi, W., Luo, M., Hu, W., Huang, Y., Jiang, F., Xie, F., & Zhang, B. (2022). Increasing xanthan gum content could enhance the performance of agar/konjac glucomannan-based system. *Food Hydrocolloids*, *132*, 107845. DOI: 10.1016/j.foodhyd.2022.107845

Ravenel, N. (2010, September). Pemulen® TR-2: An emulsifying agent with promise. *WAAC Newsletter*, *32*(3), 10–12. Available at: https://cool.culturalheritage.org/waac/wn/wn32/wn32-3/wn32-304.pdf

Reger, M., Sekine, T., & Hoffmann, H. (2012). Pickering emulsions stabilized by amphiphile covered clays. *Colloids and Surfaces A: Physicochemical and Engineering Aspects*, *413*, 25–32. DOI: 10.1016/j.colsurfa.2011.12.005

Riedo, C., Caldera, F., Poli, T., & Chiantore, O. (2015). Poly(vinylalcohol)-borate hydrogels with improved features for the cleaning of cultural heritage surfaces. *Heritage Science*, *3*(1), 23. DOI: 10.1186/s40494-015-0053-2

Rochas, C., & Lahaye, M. (1989). Average molecular weight and molecular weight distribution of agarose and agarose-type polysaccharides. *Carbohydrate Polymers*, *10*(4), 289–298. DOI: 10.1016/0144-8617(89)90068-4

Rodrigues, L. B. O., Veloso, C. M., Bonomo, R. C. F., Rodrigues, L. B., Minim, L. A., & Costa, R. A. S. (2018). Rheological and textural studies of arrowroot (Maranta arundinacea) starch–sucrose–whey protein concentrate gels. *Journal of Food Science and Technology*, *55*(8), 2974–2984. DOI: 10.1007/s13197-018-3215-x

Russick, S., Dobrowolski, N., Nosek, B., & Williams, R. (2018). Gellan gum tips. *The Book & Paper Group Annual*, *37*, 162–164. Available at: https://cool.culturalheritage.org/coolaic/sg/bpg/annual/v37/bpga37-27.pdf

São João, J., Branco, L. C., & Fragoso, S. L. (2017). Trials of agar gels and task-specific salts for the electrochemical reduction of silver sulphide on silver leaf. In L. V. Angelova, B. Ormsby, J. Townsend, & R. Wolbers (Eds.), *Gels in the Conservation of Art* (pp. 287–291). Archetype.

Sawdy, A., Lubelli, B., Voronina, V., & Pel, L. (2010). Optimizing the extraction of soluble salts from porous materials by poultices. *Studies in Conservation*, *55*(1), 26–40. DOI: 10.1179/sic.2010.55.1.26

Scott, C. L. (2012). The use of agar as a solvent gel in objects conservation. *Objects Specialty Group Postprints*, *19*, 71–83.

Scott, C. L. (2017). Getting into the (sea)weeds: The use of rigid agar gels for the cleaning of a birch bark canoe. In L. V. Angelova, B. Ormsby, J. Townsend, & R. Wolbers (Eds.), *Gels in the Conservation of Art* (pp. 145–147). Archetype.

Shashoua, Y., Alterini, M., Pastorelli, G., & Cone, L. (2021). From microfibre cloths to poly(vinyl alcohol) hydrogels – conservation cleaning of plastics heritage. *Journal of Cultural Heritage*, *52*, 38–43. DOI: 10.1016/j.culher.2021.08.009

Shinsho, A., Brenner, T., Descallar, F. B., Tashiro, Y., Ando, N., Zhou, Y., Ogawa, H., & Matsukawa, S. (2020). The thickening properties of native gellan gum are due to freeze drying–induced aggregation. *Food Hydrocolloids*, *109*, 105997. DOI: 10.1016/j.foodhyd.2020.105997

Singh, S. P., Tandon, B. N., Rana, K. S., & Ravi, K. (1983). Chemical cleaning of marble surface of Taj Mahal using absorption clay pack. *Conservation of Cultural Property in India*, 16, 87–89.

Smith, S. (2017). The removal of complex soiling on a collection of modern art bronzes using buffered pH-adjusted aqueous gels. In L. V. Angelova, B. Ormsby, J. Townsend, & R. Wolbers (Eds.), *Gels in the Conservation of Art* (pp. 349–355). Archetype.

Stagno, V., Ciccola, A., Curini, R., Postorino, P., Favero, G., & Capuani, S. (2021). Non-invasive assessment of PVA-borax hydrogel effectiveness in removing metal corrosion products on stones by portable NMR. *Gels*, *7*(4), 265. DOI: 10.3390/gels7040265

Stavroudis, C. (2010, September). Using Pemulen with the MCP. *WAAC Newsletter*, *32*(3), 16. Available at: https://cool.culturalheritage.org/waac/wn/wn32/wn32-3/wn32-307.pdf

Stavroudis, C. (2012b). Pemulen® revised: Phuck the pH Meter. *WAAC Newsletter*, *34*(2), 19. Available at: https://cool.culturalheritage.org/waac/wn/wn34/wn34-2/wn34-206.pdf

Stavroudis, C. (2016a, July 12). *The Just and True Formulation of Emulsions Using Pemulen TR2*. Cleaning of Acrylic Painted Surfaces. Available at: https://www.getty.edu/conservation/our_projects/education/caps/just_true_using_Pemulen_TR2.pdf

Stavroudis, C. (2017). Gels: Evolution in practice. In L. V. Angelova, B. Ormsby, J. Townsend, & R. Wolbers (Eds.), *Gels in the Conservation of Art* (pp. 209–217). Archetype.

Stiglitz, M., & Bearman, J. (2016). Of earth and sky: A pair of Ming hanging scrolls, from past repairs to present conservation. *Restaurator*, *37*(3–4). DOI: 10.1515/res-2016-0011

Storace, M. S., & Iannuccelli, S. (2005). The conservation of Matthaeus Greuter's ancient map of Italy. *Beiträge Zur Erhaltung von Kunst- Und Kulturgut*, *1*.

Stulik, D., & Miller, D. (2004). Research into Solvent Gel Residues. In V. Dorge (Ed.), *Solvent Gels for the Cleaning of Works of Art: The Residue Question* (pp. 18–53). Getty Publications.

Stulik, D., Miller, D., Khanjian, H., Khandekar, N., Wolbers, R., Carlson, J., & Peterson, W. C. (2004). *Solvent Gels for the Cleaning of Works of Art: The Residue Question* (V. Dorge, Ed.). Getty Publications.

Stulik, D., & Wolbers, R. (2004). Project outcome, spin-offs, and future research needs. In V. Dorge (Ed.), *Solvent Gels for the Cleaning of Works of Art: The Residue Question*. (pp. 131–144). Getty Publishing.

Sullivan, M., Brogdon-Grantham, S., & Taira, K. (2014). New approaches to cleaning works of art on paper and photographs. *Student Papers & Posters*. ANAGPIC, the Association of North American Graduate Programs in Conservation, Buffalo State College. Available at: https://faic.wpenginepowered.com/anagpic-student-papers/wp-content/uploads/sites/11/2016/02/anagpic2014_sullivan_etal_paper.pdf

Sullivan, M. R., Duncan, T. T., Berrie, B. H., & Weiss, R. G. (2017). Rigid polysaccharide gels for paper conservation: A residue study. In L. V. Angelova, B. Ormsby, J. Townsend, & R. Wolbers (Eds.), *Gels in the Conservation of Art* (pp. 42–50). Archetype.

Tako, M. (1992). Synergistic interaction between xanthan and konjac glucomannan in aqueous media. *Bioscience, Biotechnology, and Biochemistry*, *56*(8), 1188–1192. DOI: 10.1271/bbb.56.1188

Tamura, M. (2017). A shortage of agar: The sustainability of a common conservation material. *WAAC Newsletter*, *39*(1), 12–13.

Tamura, M., & Takagi, K. (2017). *Towards the Sustainable Use of Agar/Agarose in Conservation: A Case Study of the Izu Peninsula, Japan*. Archetype.

Thomas, W. R. (1997). Konjac gum. In A. P. Imeson (Ed.), *Thickening and Gelling Agents for Food* (pp. 169–179). Springer US. DOI: 10.1007/978-1-4615-2197-6_8

Tobin, G., & Sawicki, M. (2023, September 18). Developing conservation practices for cleaning gilded surfaces: Applications for xPVOH-borax organogels to clean two gilded frames. *Working Towards a Sustainable. ICOM-CC Preprints*. ICOM-CC 20th Triennial Conference, Valencia. Available at: https://www.icom-cc-publications-online.org/5600/Developing-conservation-practices-for-cleaning-gilded-surfaces--Applications-for-xPVOH-borax-organogels-to-clean-two-gilded-frames

Tomás, H., Alves, C. S., & Rodrigues, J. (2018). Laponite®: A key nanoplatform for biomedical applications? *Nanomedicine: Nanotechnology, Biology and Medicine*, *14*(7), 2407–2420. DOI: 10.1016/j.nano.2017.04.016

Tong, D. A. (2013). *The Evaluation of Modified Laponite Solvent-Gel as a Poultice in Paper Conservation*. [Poster] Queen's University, Canada.

Tortajada Hernando, S., & Blanco Domínguez, M. M. (2013). Cleaning plaster surfaces with agar-agar gels: Evaluation of the technique. *Ge-Conservación: Publicación Digital Hispano-Lusa de Conservación y Restauración*, *4*. DOI: 10.37558/gec.v4i0.153

Vallieres, J. (2013). *Gellan Gum: Investigating Applications as a Solvent Gel*. [Poster] Queen's University, Canada.

van Dyke, Y. (2004a). Practical applications of protease enzymes in paper conservation. *The Book and Paper Group Annual, 23*, 93–107.

van Dyke, Y. (2009b). Sacred Leaves: The conservation and exhibition of early Buddhist manuscripts on palm leaves. *The Book and Paper Group Annual, 28*. Available at: https://cool.culturalheritage.org/coolaic/sg/bpg/annual/v28/bpga28-17.pdf

van Dyke, Y. (2017a). Agarose-enzyme gels in paper conservation. In L. V. Angelova, B. Ormsby, J. Townsend, & R. Wolbers (Eds.), *Gels in the Conservation of Art* (pp. 101–106). Archetype.

Varges, P. R., Costa, C., Fonseca, B., Naccache, M., & De Souza Mendes, P. R. (2019). Rheological characterization of Carbopol® dispersions in water and in water/glycerol solutions. *Fluids, 4*(1), 3. DOI: 10.3390/fluids4010003

Warda, J., Brückle, I., Bezúr, A., & Kushel, D. (2007). Analysis of agarose, Carbopol, and Laponite gel poultices in paper conservation. *Journal of the American Institute for Conservation, 46*(3), 263–279. DOI: 10.1179/019713607806112260

Williams, D. (2010, September). Pemulen case study: A Midsummer-Night's Dream. *WAAC Newsletter, 32*(3), 14–15. Available at: https://cool.culturalheritage.org/waac/wn/wn32/wn32-3/wn32-306.pdf

Wolbers, R. (2017). Terminology and properties of selected Gels. In L. V. Angelova, B. Ormsby, J. Townsend, & R. Wolbers (Eds.), *Gels in the Conservation of Art* (pp. 381–394). Archetype.

Woolfitt, C., & Abrey, G. (2000). *Poultices and the cleaning of historic masonry and sculpture*. Building Conservation. [online] Available at: https://www.buildingconservation.com/articles/poultices/poultice.htm [Accessed 11 Sep. 2023]

Wu, W., Que, F., Li, X., Shi, L., Deng, W., Fu, X., Xiong, G., Sun, J., Wang, L., & Xiong, S. (2022a). Effects of enzymatic konjac glucomannan hydrolysates on textural properties, microstructure, and water distribution of grass carp surimi gels. *Foods, 11*(5), 750. DOI: 10.3390/foods11050750

Zeece, M. (2020). Chapter Seven - Food additives. In M. Zeece (Ed.), *Introduction to the Chemistry of Food* (pp. 251–311). Academic Press. DOI: 10.1016/B978-0-12-809434-1.00007-4

Zeng, M., Morris, C. F., Batey, I. L., & Wrigley, C. W. (1997b). Sources of variation for starch gelatinization, pasting, and gelation properties in wheat. *Cereal Chemistry, 74*(1), 63–71. DOI: 10.1094/CCHEM.1997.74.1.63

## 4 – Cleaning

### *Enzymes*

Abdelaal, S., & Sandu, I. C. A. (2019). Assessment of protease in cleaning of bat blood patches from ancient Egyptian wall paintings and surface inscriptions. *International Journal of Conservation Science, 10*(3), 459–474. Available at: https://ijcs.ro/public/IJCS-19-42_Abdelaal.pdf

Abdel-Nasser, M., Abdel-Maksoud, G., Abdel-Aziz, M. S., Darwish, S. S., Hamed, A. A., & Youssef, A. M. (2022). Evaluation of the efficiency of nanoparticles for increasing α-amylase enzyme activity for removing starch stain from paper artifacts. *Journal of Cultural Heritage, 53*, 14–23. DOI: 10.1016/j.culher.2021.11.004

Aehle, W., Perham, R. N., Michal, G., Caddow, A. J., & Concoby, B. (2008). Enzymes, 1. General. In *Ullmann's Encyclopedia of Industrial Chemistry*. Wiley-VCH Verlag GmbH & Co. KGaA. DOI: 10.1002/14356007.a09_341.pub3

Ahmed, H. E., Gremos, S. S., & Kolisis, F. N. (2010). Enzymatic removal of the oily dirt from a Coptic tunic using the enzyme lipase. *Journal of Textile and Apparel, Technology and Management, 6*(3). Available at: https://jtatm.textiles.ncsu.edu/index.php/JTATM/article/view/674

Ahmed, H. E., & Kolisis, F. N. (2011). An investigation into the removal of starch paste adhesives from historical textiles by using the enzyme α-amylase. *Journal of Cultural Heritage*, *12*(2), 169–179. DOI: 10.1016/j.culher.2010.08.001

Ahmed, H. E., & Kolisis, F. N. (2012). A study on using of protease for removal of animal glue adhesive in textile conservation. *Journal of Applied Polymer Science*, *124*(5), 3565–3576. DOI: 10.1002/app.34053

Andrews, T. M., Andrews, W. W., & Baker, C. (1992). An investigation into the removal of enzymes from paper following conservation treatment. *Journal of the American Institute for Conservation*, *31*(3), 313–323. DOI: 10.1179/019713692806066547

Aschengreen, N. H. (1975). Proteolytic enzymes used for restoring. *IADA (International Association of Book and Paper Conservators) Congress 1975 Copenhagen*, 97–104.

Banik, G., & Cremonesi, P. (2003). Removal of starch paste adhesives and relinings from paper-based objects by means of enzyme poultices. *Materiali Tradizionali Ed Innovativi Nella Pulitura Dei Dipinti e Delle Opere Policrome Mobili: Atti Del Convegno, Piazzola Su Brenta (PD)*, 33–38.

Barbabietola, N., Tasso, F., Alisi, C., Marconi, P., Perito, B., Pasquariello, G., & Sprocati, A. R. (2016). A safe microbe-based procedure for a gentle removal of aged animal glues from ancient paper. *International Biodeterioration & Biodegradation*, *109*, 53–60. DOI: 10.1016/J.IBIOD.2015.12.019

Bartl, B., Kotlík, P., Ďurovič, M., & Bacílková, B. (2011). Application of α-amylase in combination with ultrasound to remove starch based adhesives from paper. *Restaurator*, *32*(1). DOI: 10.1515/rest.2011.005

Bartoletti, A., Maor, T., Chelazzi, D., Bonelli, N., Baglioni, P., Angelova, L. V., & Ormsby, B. A. (2020). Facilitating the conservation treatment of Eva Hesse's Addendum through practice-based research, including a comparative evaluation of novel cleaning systems. *Heritage Science*, *8*(1), 35. DOI: 10.1186/s40494-020-00378-z

Bellucci, R., Cremonesi, P., & Pignagnoli, G. (1999). A preliminary note on the use of enzymes in conservation: The removal of aged acrylic resin coatings with lipase. *Studies in Conservation*, *44*(4), 278–281. DOI: 10.1179/sic.1999.44.4.278

Blüher, A., Banik, G., Maurer, K.-H., & Thobois, E. (1996b). The application of enzyme-containing methylcellulose gels for the removal of starch-based adhesives in albums. In J. Bridgland (Ed.), *ICOM Committee for Conservation 11th Triennial Meeting Edinburgh 1-6 September 1996*. Available at: https://www.icom-cc-publications-online.org/2990/The-Application-of-Enzyme-containing-Methylcellulose-Gels-for-the-Removal-of-Starch-based-Adhesives-in-Albums

Blüher, A., Grube, A., Bornscheuer, U., & Banik, G. (1997). A reappraisal of the enzyme lipase for removing drying-oil stains on paper. *The Paper Conservator*, *21*(1), 37–47. DOI: 10.1080/03094227.1997.9638597

Borring, N. J., & Skov Hilby, C. (2015). Enzymes in conservation: Separating the Triumphal Arch by Albrecht Dürer with α-amylase embedded in agarose gels. *IADA Congress XIII*. DOI: 10.13140/RG.2.1.1505.1609

Borring, N. J., Singer, H., Skov Hilby, C., & Klyver, L. (2015). Dürer under the knife! conservation of The Arch Of Honour Of Maximilian. In N. J. Borring & H. Kolind Poulsen (Eds.), *Power and Glory – Albrecht Dürer in the Emperor's Service* (pp. 82–92). Statens Museum for Kunst.

Bosch Roig, P., Regidor Ros, J. L., & Montes Estellés, R. (2013). Biocleaning of nitrate alterations on wall paintings by Pseudomonas stutzeri. *International Biodeterioration & Biodegradation*, *84*, 266–274. DOI: 10.1016/j.ibiod.2012.09.009

Bosch-Roig, P., Lustrato, G., Zanardini, E., & Ranalli, G. (2015). Biocleaning of cultural heritage stone surfaces and frescoes: Which delivery system can be the most appropriate? *Annals of Microbiology*, *65*(3), 1227–1241. DOI: 10.1007/s13213-014-0938-4

Bosch-Roig, P., & Ranalli, G. (2014). The safety of biocleaning technologies for cultural heritage. *Frontiers in Microbiology*, *5*(APR). DOI: 10.3389/fmicb.2014.00155

Bosch-Roig, P., Regidor Ros, J. L., Soriano Sancho, M. P., Montes Estellés, R., & Roig Picazo, P. (2017). Biocleaning of wall paintings on uneven surfaces with warm agar gels. In L. V. Angelova, B. Ormsby, J. Townsend, & R. Wolbers (Eds.), *Gels in the Conservation of Art* (pp. 119–121). Archetype.

Brockman, M., & Farek, E. (2018a). *Investigation into reduction of foxing stains in paper*. the Association of North American Graduate Programs in Conservation (ANAGPIC), Queens University, Kingston, Ontario.

Burns, P., & Meadow, R. H. (2013). The use of trypsin to prepare skeletal material for comparative collections with a focus on fish. *Archaeofauna*, *22*, 29–36. DOI: 10.15366/archaeofauna2013.22.002

Cappitelli, F., Toniolo, L., Sansonetti, A., Gulotta, D., Ranalli, G., Zanardini, E., & Sorlini, C. (2007). Advantages of using microbial technology over traditional chemical technology in removal of black crusts from stone surfaces of historical monuments. *Applied and Environmental Microbiology*, *73*(17), 5671–5675. DOI: 10.1128/AEM.00394-07

Cremonesi, P., & Casoli, A. (2021). Enzymes as tools for conservation of works of art. *Journal of Cultural Heritage*, *50*, 73–87. DOI: 10.1016/j.culher.2021.06.005

Decoux, S. (2002). Enzymes used for adhesive removal in paper conservation: A literature review. *Journal of the Society of Archivists*, *23*(2), 187–195. DOI: 10.1080/0033798102200000 6372

DeSantis, P. C. (1983). Some observations on the use of enzymes in paper conservation. *Journal of the American Institute for Conservation*, *23*(1), 7–27.

Erickson, H. M. (1992). Usage recommendations for alpha-amylases: Maximizing enzyme activity while minimizing enzyme-artifact binding residues. *The Book and Paper Group Annual*, 11, 24–33.

Ferrari, M., Mazzoli, R., Morales, S., Fedi, M., Liccioli, L., Piccirillo, A., Cavaleri, T., Oliva, C., Gallo, P., Borla, M., Cardinali, M., & Pessione, E. (2017). Enzymatic laundry for old clothes: Immobilized alpha-amylase from Bacillus sp. for the biocleaning of an ancient Coptic tunic. *Applied Microbiology and Biotechnology*, *101*(18), 7041–7052. DOI: 10.1007/s00253-017-8437-8

Galinsky, E., Nouaimi-Bachmann, M., Scheu, U., Bisswanger, H., Haberditzl, A. (2005). Application of immobilized enzymes for the paper-splitting process. *ICOM Committee for Conservation. 14th Triennial Meeting*, The Hague.

Gallo, C., Rizzo, P., & Guerra, G. (2020). Antimicrobial release from cleaning poultices for the conservation and disinfection of stone surfaces. *Applied Clay Science*, *193*, 105667. DOI: 10.1016/j.clay.2020.105667

Gänsicke, S., Hatchfield, P., Hykin, A., Svoboda, M., & Tsu, C. M.-A. (2003a). The ancient Egyptian collection at the Museum of Fine Arts, Boston. Part 2, a review of former treatments at the MFA and their consequences. *Journal of the American Institute for Conservation*, *42*(2), 193–245. DOI: 10.2307/3180070

Garner, R., & Horie, C. (1984). The conservation and restoration of microscope slides of mosses. *Studies in Conservation*, *29*.

Gherardi, F., Turyanska, L., Ferrari, E., Weston, N., Fay, M. W., & Colston, B. J. (2019). Immobilized enzymes on gold nanoparticles: From enhanced stability to cleaning of heritage textiles. *ACS Applied Bio Materials*, *2*(11), 5136–5143. DOI: 10.1021/acsabm.9b00802

Glenn, S., Hackett, J., Haldane, E.-A., & Im, S. (2015b). Borrowing from the neighbours: Using the technology of other disciplines to treat difficult textile conservation problems. In A. Fairhurst (Ed.), *Learning Curve: Education, Experience, Reflection. Postprints* (pp. 33–42). ICON. Available at:https://www.icon.org.uk/static/dd33714f-f5b2-4f69-8b1eb6d866d4a8b0/2015learningcurveforum.pdf

Graedel, T. E., Franey, J. P., & Kammlott, G. W. (1983a). The corrosion of copper by atmospheric sulphurous gases. *Corrosion Science*, *23*, 1141–1152.
Grattan, D. W., St. Hilaire, J., Burgess, H. D., & Maccawley, J. (1987). The characterization of enzymes for use in paper conservation. In *International Conference on the Conservation of Library and Archive Materials and the Graphic Arts, Cambridge, 22–26 September 1980. Abstracts and preprints*. Institute for Paper Conservation.
Guo, J., Shi, E., Mu, X., Han, J., Qu, W., Song, Y., Da, S., Dawa, O., Gao, F., Guo, L., & Nevin, A. (2023). Optimization of conditions for the cleaning of overpaint from the stone carvings of the Potala Palace, Tibet using enzyme-based cleaning agents. *Journal of Cultural Heritage*, *59*, 83–92. DOI: 10.1016/j.culher.2022.11.009
Hamed, S. A. E.-K. M. (2012). A preliminary study on using enzymes in cleaning archaeological wood. *Journal of Archaeological Science*, *39*(7). DOI: 10.1016/j.jas.2012.02.026
Hamed, S. A. E.-K. M. (2019). Capabilities and limitations of microbial enzymes in wood cleaning. *Mediterranean Archaeology & Archaeometry*, *19*(2). DOI: 10.5281/zenodo.3065999
Isola, D., Bartoli, F., Meloni, P., Caneva, G., & Zucconi, L. (2022). Black fungi and stone heritage conservation: Ecological and metabolic assays for evaluating colonization potential and responses to traditional biocides. *Applied Sciences (Switzerland)*, *12*(4). DOI: 10.3390/app12042038
Jia, Y., Sciutto, G., Botteon, A., Conti, C., Focarete, M. L., Gualandi, C., Samorì, C., Prati, S., & Mazzeo, R. (2021b). Deep eutectic solvent and agar: A new green gel to remove proteinaceous-based varnishes from paintings. *Journal of Cultural Heritage*, *51*, 138–144. DOI: 10.1016/j.culher.2021.08.001
Joseph, E. (Ed.). (2021). *Microorganisms in the Deterioration and Preservation of Cultural Heritage*. Springer International Publishing. DOI: 10.1007/978-3-030-69411-1
Kavkler, K., Gunde-Cimerman, N., Zalar, P., & Demšar, A. (2015). Fungal contamination of textile objects preserved in Slovene museums and religious institutions. *International Biodeterioration & Biodegradation*, *97*. DOI: 10.1016/j.ibiod.2014.09.020
Know, S. (1995). An assessment of catalytic activity in Wolbers's lipase gel using an enzyme assay. *Conservation News (United Kingdom Institute for Conservation of Historic and Artistic Works)*, *58*.
Kouznetsov, D. A. (2000). Biochemical methods in cultural heritage conservation studies: An alkylation enzyme, s-adenosylmethionine transmethylase. *Studies in Conservation*, *45*(2), 117–126. DOI: 10.1179/sic.2000.45.2.117
Larminie, P. (1992). The use of enzymes for the removal of oil stains from paper. In *Institute of Paper Conservation: Conference papers, Manchester, 1992*. Institute of Paper Conservation.
Lee, C., Di Turo, F., Vigani, B., Weththimuni, M. L., Rossi, S., Beltram, F., Pingue, P., Licchelli, M., Malagodi, M., Fiocco, G., & Volpi, F. (2022b). Biopolymer gels as a cleaning system for differently featured wooden surfaces. *Polymers*, *15*(1), 36. DOI: 10.3390/polym15010036
Lee, J., Xiang, L., Byambabaatar, S., Kim, H., Jin, K. S., & Ree, M. (2019). Bacillus licheniformis α-amylase: Structural feature in a biomimetic solution and structural changes in extrinsic conditions. *International Journal of Biological Macromolecules*, *127*, 286–296. DOI: 10.1016/J.IJBIOMAC.2019.01.053
Markevičius, T., Syversen, T., Chan, E., Olsson, N., Hilby, C. S., & Šimaitė, R. (2017b). Cold, warm, warmer: Use of precision heat transfer in the optimization of hydrolytic enzyme and hydrogel cleaning systems. In L. V. Angelova, B. Ormsby, J. Townsend, & R. Wolbers (Eds.), *Gels in the Conservation of Art* (pp. 68–72). Archetype.

Mazzoli, R., Giuffrida, M. G., & Pessione, E. (2018). Back to the past—forever young: Cutting-edge biochemical and microbiological tools for cultural heritage conservation. *Applied Microbiology and Biotechnology*, *102*(16), 6815–6825. DOI: 10.1007/s00253-018-9121-3

Mazzuca, C., Micheli, L., Cervelli, E., Basoli, F., Cencetti, C., Coviello, T., Iannuccelli, S., Sotgiu, S., & Palleschi, A. (2014b). Cleaning of paper artworks: Development of an efficient gel-based material able to remove starch paste. *ACS Applied Materials and Interfaces*, *6*(19), 16519–16528. DOI: 10.1021/am504295n

McDonald, A. G., & Tipton, K. F. (2023). Enzyme nomenclature and classification: The state of the art. *The FEBS Journal*, *290*(9), 2214–2231. DOI: 10.1111/FEBS.16274

Mina, L. (2020b). Foxy underpants: Or the use of chelators and enzymes to reduce foxing stains on early nineteenth century men's linen underpants. *Journal of the American Institute for Conservation*, *59*(1), 3–17. DOI: 10.1080/01971360.2019.1674604

Nielsen, M. H., Jepsen, S. J., & Outtrup, H. (1981). Enzymes for low temperature washing. *Journal of the American Oil Chemists' Society*, *58*(5), 644–649. DOI: 10.1007/BF02672384

Nikolova, D. (1993). Proteinase inhibitors from vegetables and their application in enzymatic conservation treatments. *Restaurator*, *14*(4). DOI: 10.1515/rest.1993.14.4.199

Ntasi, G., Sbriglia, S., Pitocchi, R., Vinciguerra, R., Melchiorre, C., Dello Ioio, L., Fatigati, G., Crisci, E., Bonaduce, I., Carpentieri, A., Marino, G., & Birolo, L. (2022b). Proteomic characterization of collagen-based animal glues for restoration. *Journal of Proteome Research*. DOI: 10.1021/acs.jproteome.2c00232

Okamoto, I., & Minagawa, M. (1987). Removal of protein stains from fabrics–presoaking with protease–effect of mechanical action for detergency and damage of cotton fabrics soiled with epidermal stratum corneum debris. *Journal of the Japan Research Association for Textile End-Uses / Sen'i seihin shohi kagaku*, *28*(12). DOI: 10.11419/senshoshi1960.28.522

Okamoto, I., & Minagawwa, M. (1985). Studies on the removal of protein stains: Removal of epidermal stratum corneum debris adhered to various fabrics in dry cleaning system. *Journal of the Japan Research Association for Textile End-Uses / Sen'i seihin shohi kagaku*, *26*(12). DOI: 10.11419/senshoshi1960.26.511

Palla, F. (2013). Bioactive molecules: Innovative contributions of biotechnology to the restoration of cultural heritage. *Conservation Science in Cultural Heritage*, *13*, 369–373. DOI: 10.6092/ISSN.1973-9494/4191

Palla, F., Barresi, G., Giordano, A., Schiavone, S., Trapani, M. R., Rotolo, V., Parisi, M. G., & Cammarata, M. (2016). Cold-active molecules for a sustainable preservation and restoration of historic-artistic manufacts. *International Journal of Conservation Science*, *7*(SI1), 239–246.

Pereira, C., Busani, T., Branco, L. C., Joosten, I., & Sandu, I. C. A. (2013). Nondestructive characterization and enzyme cleaning of painted surfaces: Assessment from the macro to nano level. *Microscopy and Microanalysis*, *19*(6), 1632–1644. DOI: 10.1017/S1431927613013196

Petrella, G., Mazzuca, C., Micheli, L., Cervelli, E., De Fazio, D., Iannuccelli, S., Sotgiu, S., Palleschi, G., & Palleschi, A. (2016b). A new sustainable and innovative work for paper artworks cleaning process: Gellan hydrogel combined with hydrolytic enzymes. *International Journal of Conservation Science*, *7*(special 1), 273–280.

Rivers, S., & Umney, N. (2013b). *Conservation of Furniture*. Routledge.

Ruginescu, R., Enache, M., Popescu, O., Gomoiu, I., Cojoc, R., Batrinescu-Moteau, C., Maria, G., Dumbravician, M., & Neagu, S. (2022). Characterization of some salt-tolerant bacterial hydrolases with potential utility in cultural heritage bio-cleaning. *Microorganisms*, *10*(3). DOI: 10.3390/microorganisms10030644

Salvador, C., Sandu, I., Sandbakken, E., Candeias, A., & Caldeira, A. T. (2022). Biodeterioration in art: A case study of Munch's paintings. *European Physical Journal Plus*, *137*(1). DOI: 10.1140/epjp/s13360-021-02187-0

Sar, P., Kazy, S. K., Asthana, R. K., & Singh, S. P. (1999). Metal adsorption and desorption by lyophilized Pseudomonas aeruginosa. *International Biodeterioration & Biodegradation, 44*(2–3). DOI: 10.1016/S0964-8305(99)00064-5

Sato, M., Park, U. K., & Minagawa, M. (1989). Studies on the enzyme-containing detergents for foam washing. Part 2, Effect of the stability of protease activity on the removal of gelatin from cotton fabric by foam. *Journal of the Japan Research Association for Textile End-Uses / Sen'i seihin shohi kagaku, 30*(7). DOI: 10.11419/senshoshi1960.30.317

Schröer, L., Fiers, G., Deprez, M., Boon, N., Cnudde, V., Soens, L., & De Kock, T. (2022a). Examining the potential of enzyme-based detergents to remove biofouling from limestone heritage. *Coatings, 12*(3). DOI: 10.3390/coatings12030375

Schwarz, I., Blüher, A., Banik, G., Thobois, E., & Maurer, K.-H. (1999). The development of a ready-for-use poultice for local removal of starch paste by enzymatic action. *Restaurator*, 225–244. DOI: 10.1515/rest.1999.20.3-4.225

Segal, J. (1994). New techniques for the application of enzymes. In *Conservation of Historic and Artistic Works on Paper: Proceedings of a Conference, Ottawa, October 3-7, 1988*. Canadian Conservation Institute.

Segal, J., & Cooper, D. (1977). The use of enzymes to release adhesives. *The Paper Conservator, 2*.

Sharma, A., Gautam, S., & Wadhawan, S. (2014). Xanthomonas. In C. A. Batt & M. L. Tortorello (Eds.), *Encyclopedia of Food Microbiology* (2nd Edition; pp. 811–817). Academic Press. DOI: 10.1016/B978-0-12-384730-0.00359-1

Smith, L. (1987). Use of enzymes in textile cleaning. In Preprints of papers presented at *the fifteenth annual meeting of the American Institute for Conservation of Historic and Artistic Works. Vancouver, British Columbia, Canada, May 20-24, 1987*. American Institute for Conservation of Historic and Artistic Works.

Suwannaphan, S., Fufeungsombut, E., Promboon, A., & Chim-Anage, P. (2017). A serine protease from newly isolated Bacillus sp. for efficient silk degumming, sericin degrading and colour bleaching activities. *International Biodeterioration and Biodegradation, 117*, 141–149. DOI: 10.1016/j.ibiod.2016.12.009

Taniguchi, H., & Honnda, Y. (2009). Amylases. In M. Schaechter (Ed.), *Encyclopedia of Microbiology* (3rd Edition) (pp. 159–173). Academic Press. DOI: 10.1016/B978-012373944-5.00130-9

Tímár-Balázsy, Á., & Eastop, D. (1998a). *Chemical Principles of Textile Conservation*. Butterworth-Heinemann.

Tokoro, Y., & Minagawa, M. (1985). Studies on the removal of blood protein stains from fabrics. Part 4, Influences of pH and temperature characteristics of protease on the removal of blood protein stains. *Journal of the Japan Research Association for Textile End-Uses / Sen'i seihin shohi kagaku, 26*(11). DOI: 10.11419/senshoshi1960.26.479

Tortora, M., Gherardi, F., Ferrari, E., & Colston, B. (2020). Biocleaning of starch glues from textiles by means of α-amylase-based treatments. *Applied Microbiology and Biotechnology, 104*(12), 5361–5370. DOI: 10.1007/s00253-020-10625-9

van Dyke, Y. (2004b). Practical applications of protease enzymes in paper conservation. *The Book and Paper Group Annual, 23*, 93–107. http://aic.stanford.edu/bpg/annual/

van Dyke, Y. (2009c). Sacred Leaves: The conservation and exhibition of early Buddhist manuscripts on palm leaves. *The Book and Paper Group Annual, 28*. https://cool.culturalheritage.org/coolaic/sg/bpg/annual/v28/bpga28-17.pdf

van Dyke, Y. (2017b). Agarose-enzyme gels in paper conservation. In L. V. Angelova, B. Ormsby, J. Townsend, & R. Wolbers (Eds.), *Gels in the Conservation of Art* (pp. 101–106). Archetype.

Vokic, D., & Berovič, M. (2005, September 12). Use of lipase to remove oil-based overpaints. *ICOM Committee for Conservation*. 14th Triennial Meeting, The Hague. https://www.icom-cc-publications-online.org/2144/Use-of-lipase-to-remove-oil-based-overpaints

von Endt, D. W., Ross, C. A., & Hare, P. E. (1999). Initial results from cleaning small vertebrate skeletons using the enzyme trypsin. *Collection Forum*, *13*(2), 51–62.
Wu, W., Que, F., Li, X., Shi, L., Deng, W., Fu, X., Xiong, G., Sun, J., Wang, L., & Xiong, S. (2022b). Effects of enzymatic konjac glucomannan hydrolysates on textural properties, microstructure, and water distribution of grass carp surimi gels. *Foods*, *11*(5), 750. DOI: 10.3390/foods11050750

### *pH, Buffers, and Chelation*

Adler, S., & Eaton, L. (1995). Chelating agents in wet cleaning systems. *AIC Textile Specialty Group Postprints*, 69–78. Available at: http://www.museumtextiles.com/uploads/7/8/9/0/7890082/chelating_agents_1995_aic_tsg_postprints.pdf
Basolo, F., Chen, Y. T., & Murmann, R. K. (1954). Steric effects and the stability of complex compounds. IV. the chelating tendencies of c-substituted ethylenediamines with copper(II) and nickel(II) ions1,2. *Journal of the American Chemical Society*, *76*(4), 956–959. DOI: 10.1021/ja01633a002
Bazemore, A. (2017). Chelating soluble iron (II) from iron gall ink using calcium phytate in agar gel. In L. V. Angelova, B. Ormsby, J. Townsend, & R. Wolbers (Eds.), *Gels in the Conservation of Art* (pp. 116–118). Archetype.
Berlangieri, C., Andrina, E., Matarrese, C., Carretti, E., Traversi, R., Severi, M., Chelazzi, D., Dei, L., & Baglioni, P. (2017b). Chelators confined into 80pvac-borax highly viscous dispersions for the removal of gypsum degradation layers. *Pure and Applied Chemistry*, *89*(1), 97–109. DOI: 10.1515/pac-2016-0815
Brockman, M. (2020). Accelerated aging study of papers treated with citrate solutions. *WAAC Newsletter*, *42*(1), 1[illegible]–13.
Brockman, M., & Farek, E. (2018b). *Investigation into reduction of foxing stains in paper*. The Association of North American Graduate Programs in Conservation (ANAGPIC), Queens University, Kingston, Ontario.
Burgos-Cara, A., Putnis, C. V., Ortega-Huertas, M., & Ruiz-Agudo, E. (2017). Influence of pH and citrate on the formation of oxalate layers on calcite revealed by in situ nanoscale imaging. *CrystEngComm*, *19*(25), 3420–3429. DOI: 10.1039/C7CE00305F
Burnstock, A., & van den Berg, K. J. (2005). A pilot study of the effects of triammonium citrate solutions used for surface cleaning paintings. *Postprints of the American Institute for Conservation of Historic and Artistic Works*. vol. 17, Paintings Speciality Group, pp. 56–64
Carlyle, L., Townsend, J., & Hackney, S. (n.d.). Triammonium citrate an investigation into its application for surface cleaning. In S. Hackney, J. Townsend, & N. Eastaugh (Eds.), *Dirt and Pictures Separated: Papers Given at a Conference Held Jointly by UKIC and the Tate Gallery, January 1990* (pp. 44–48). United Kingdom Institute of Conservation.
Chiampo, F., & Zacchini, M. (2021). Environmental restoration of metal-contaminated soils. *Applied Sciences*, *11*(22), 10805. DOI: 10.3390/app112210805
Cohen, G., Lewis, D., & Sinet, P. M. (1981). Oxygen consumption during the fenton-type reaction between hydrogen peroxide and a ferrous-chelate (Fe2+-DTPA). *Journal of Inorganic Biochemistry*, *15*(2), 143–151. DOI: 10.1016/S0162-0134(00)80298-6
Crisci, G. M., La Russa, M. F., Macchione, M., Malagodi, M., Palermo, A. M., & Ruffolo, S. A. (2010). Study of archaeological underwater finds: Deterioration and conservation. *Applied Physics. A, Materials Science & Processing*, *100*(3). DOI: 10.1007/s00339-010-5661-9
Dabrowa, B. (2004). The conservation of three gilded frames for the new paintings galleries at the Victoria and Albert Museum. *V & A Conservation Journal*, *46*. Available at: http://www.vam.ac.uk/content/journals/conservation-journal/issue-46/the-conservation-of-three-gilded-frames-for-the-new-paintings-galleries-at-the-victoria-and-albert-museum/

Dow Chemical Company. (2021). *Chelation chemistry: General concepts of the chemistry of chelation* (Nos. 113-01388-01-0721 S2D). https://www.dow.com/content/dam/dcc/documents/en-us/app-tech-guide/113/113-01388-01-chelation-chemistry-general-concepts-of-the-chemistry-of-chelation.pdf

Dwan, A. (2015). Ammonium citrates for stain removal in paper. *Newsletter (Western Association for Art Conservation), 37*(3). Available at: https://www.waac-us.org/_files/ugd/d3b1ca_176ea28a256148f0b46d6c8d8eb6b4de.pdf

El-Gohary, M., & Saad, M. (2020a). Using some traditional and nanomaterials in the cleaning of ancient Egyptian pottery artifacts (jar and shallow plate) from Athribis. *Archaeological and Anthropological Sciences, 12*(12). DOI: 10.1007/s12520-020-01235-7

Funke, A., Poole, L., Church, J., Striegel, M., & Singer, M. (2017). The use of medical chelating agents for the removal of iron stains from marble. *Objects Specialty Group Postprints, 24*, 235–249.

Gänsicke, S., Hatchfield, P., Hykin, A., Svoboda, M., & Tsu, C. M.-A. (2003b). The ancient Egyptian collection at the Museum of Fine Arts, Boston. Part 2, a Review of Former Treatments at the MFA and Their Consequences. *Journal of the American Institute for Conservation, 42*(2), 193–245. DOI: 10.2307/3180070

Gent, M., & Rees, J. (1994). A conservation treatment to remove residual iron from platinum prints. *The Paper Conservator, 18*, 90–95.

Gervais, C., Grissom, C. A., Little, N., & Wachowiak, M. J. (2010). Cleaning marble with ammonium citrate. *Studies in Conservation, 55*(3), 164–176. DOI: 10.1179/sic.2010.55.3.164

Godfrey, I., Kasi, K., & Richards, V. (2001, June 11). Iron removal from waterlogged leather and rope recovered from shipwreck sites. *Proceedings of the 8th ICOM Group on Wet Organic Archaeological Materials Conference.*

Graedel, T. E., Franey, J. P., & Kammlott, G. W. (1983b). The corrosion of copper by atmospheric sulphurous gases. *Corrosion Science, 23*, 1141–1152.

Hart, J. R. (2011). Ethylenediaminetetraacetic acid and related chelating agents. In *Ullmann's Encyclopedia of Industrial Chemistry*. Wiley-VCH Verlag GmbH & Co. KGaA. DOI: 10.1002/14356007.a10_095.pub2

Loughead, I. S. (1998). *The Removal of Iron Corrosion from Tinplate Using Gelatin Poultices*. Queen's University.

Mardikian, P., Baatz, J. E., de Viviés, P., Drews, M. J., Grieve, S., Nicot, F. (2005). A preliminary investigation into the effects of diammonium citrate on waterlogged leather from the H.L Hunley submarine. *Proceedings of the 9th ICOM-CC Group on Wet Organic Archaeological Materials Conference*, 513–529.

Metheny, N. A., Gunn, E. M., Rubbelke, C. S., Quillen, T. F., Ezekiel, U. R., & Meert, K. L. (2017b). Effect of pH test-strip characteristics on accuracy of readings. *Critical Care Nurse, 37*(3), 50–58. DOI: 10.4037/ccn2017199

Mina, L. (2020c). Foxy underpants: Or the use of chelators and enzymes to reduce foxing stains on early nineteenth century men's linen underpants. *Journal of the American Institute for Conservation, 59*(1), 3–17. DOI: 10.1080/01971360.2019.1674604

Morrison, R., Bagley-Young, A., Burnstock, A., Van Den Berg, K. J., & Van Keulen, H. (2007). An investigation of parameters for the use of citrate solutions for surface cleaning unvarnished paintings. *Studies in Conservation, 52*(4), 255–270. DOI: 10.1179/sic.2007.52.4.255

Motekaitis, R. J., Cox Iii, X. B., Taylor, P., Martell, A. E., Miles, B., & Tvedt Jr, T. J. (1982). Thermal degradation of EDTA chelates in aqueous solution. *Canadian Journal of Chemistry, 60*(10), 1207–1213. DOI: 10.1139/v82-179

Nucera, A., Carniato, F., Baranyai, Z., Platas-Iglesias, C., & Botta, M. (2022). Characterization of the Fe(III)-Tiron system in solution through an integrated approach combining NMR relaxometric, thermodynamic, kinetic, and computational data. *Inorganic Chemistry*. DOI: 10.1021/acs.inorgchem.2c04393

Phenix, A., & Burnstock, A. (1992). The removal of surface dirt on paintings with chelating agents. *The Conservator, 16*. DOI: 10.1080/01400096.1992.9635624
Plenderleith, H. J. (1956). *The Conservation of Antiquities and Works of Art: Treatment, Repair, and Restoration*. Oxford University Press.
Frestowitz, B., Theodorakopoulos, C., & Colbourne, J. (2016). Preliminary investigation into the use of diethylenetriaminepentaacetic acid and ethylenediaminetetraacetic acid to treat iron induced foxing in paper objects. *The Book and Paper Group Annual, 35*, 163–170.
Rapti, S., Boyatzis, S., Rivers, S., Velios, A., & Pournou, A. (2017). Removing iron stains from wood and textile objects: Assessing gelled siderophores as novel green chelators. In L. V. Angelova, B. Ormsby, J. Townsend, & R. Wolbers (Eds.), *Gels in the Conservation of Art* (pp. 343–348). Archetype.
Richey, W. D. (1975). Chelating agents – a review. *Studies in Conservation, 20*(sup1), 229–234. DOI: 10.1179/sic.1975.s1.038
Rivers, S., & Umney, N. (2013c). *Conservation of Furniture*. Routledge.
Spile, S., Suzuki, T., Bendix, J., & Simonsen, K. P. (2016a). Effective cleaning of copper stained calcareous stone. *Heritage Science, 4*(1), 29. DOI: 10.1186/s40494-016-0102-5
Spile, S., Suzuki, T., Bendix, J., & Simonsen, K. P. (2016b). Effective cleaning of rust stained marble. *Heritage Science, 4*(1), 12. DOI: 10.1186/s40494-016-0081-6
Stambolov, T. (1968). Notes on the removal of iron stains from calcareous stone. *Studies in Conservation 13*(1), 45–47.
Stavroudis, C. (1985). DTPA, a chelate that helps two ways. *The Abbey Newsletter, 9*(5), 80. AATA. https://cool.culturalheritage.org/byorg/abbey/an/an09/an09-5/an09-505.html
Stavroudis, C., & Dwan, A. (2020, May). Chelating agents in paper conservation. *WAAC Newsletter, 42*(2), 16–22. https://170517c7-5797-4867-b76e-a5e55f7532ec.filesusr.com/ugd/d3b1ca_36b1fd51da424cdfba7beb10832c5ebe.pdf
Thorn, A. (1993). The impact of disodium EDTA on stone. *ICOM Committee for Conservation Tenth Triennial Meeting*, 96, 357–363.
Verhoff, F. H., & Bauweleers, H. (2014). Citric Acid. In *Ullmann's Encyclopedia of Industrial Chemistry* (pp. 1–11). Wiley-VCH Verlag GmbH & Co. KGaA. DOI: 10.1002/14356007.a07_103.pub3
Wolbers, R. (2000a). *Cleaning Painted Surfaces: Aqueous Methods*. Archetype.

### *Surfactants and emulsification*

Abbott, S. (2019). *Practical Surfactants*. Prof Steven Abbott. [online] Available at: https://www.stevenabbott.co.uk/practical-surfactants/basics.php [Accessed 24 Jan. 2024]
Baglioni, M., Montis, C., Brandi, F., Guaragnone, T., Meazzini, I., Baglioni, P., & Berti, D. (2017). Dewetting acrylic polymer films with water/propylene carbonate/surfactant mixtures - Implications for cultural heritage conservation. *Physical Chemistry Chemical Physics, 19*(35), 23723–23732. DOI: 10.1039/c7cp02608k
Baglioni, M., Poggi, G., Chelazzi, D., & Baglioni, P. (2021). Advanced materials in cultural heritage conservation. *Molecules, 26*(13). DOI: 10.3390/molecules26133967
Baglioni, P., Berti, D., Bonini, M., Carretti, E., Dei, L., Fratini, E., & Giorgi, R. (2014). Micelle, microemulsions, and gels for the conservation of cultural heritage. *Advances in Colloid and Interface Science, 205*, 361–371. DOI: 10.1016/j.cis.2013.09.008
Berti, D. (2013). Cleaning III: Emulsions and Microemulsions. In *Nanoscience for the Conservation of Works of Art*. Royal Society of Chemistry. Available at: https://books.rsc.org/books/edited-volume/1015/chapter-abstract/812652/Cleaning-III-Emulsions-and-Microemulsions?redirectedFrom=fulltext
El-Gohary, M., & Saad, M. (2020b). Using some traditional and nanomaterials in the cleaning of ancient Egyptian pottery artifacts (jar and shallow plate) from Athribis. *Archaeological and Anthropological Sciences, 12*(12). DOI: 10.1007/s12520-020-01235-7

Fields, J. A., Wingham, A., Hartog, F., & Daniels, V. (2004). Finding substitute surfactants for Synperonic N. *Journal of the American Institute for Conservation*, *43*(1), 55–73. DOI: 10.1179/019713604806112579

Getty Conservation Institute. (2013a, October 31). *Preparing a Dow Mineral Spirits Microemulsion (With Cosurfactants)*. [online video] Available at: https://www.youtube.com/watch?v=SGkf3i7rnDw

Getty Conservation Institute. (2013b, October 31). *Preparing a Pemulen Gel from MCP and Making an Emulsion*. [online video] Available at: https://www.youtube.com/watch?v=2O5pYyc45Qo

Gradzielski, M., Duvail, M., de Molina, P. M., Simon, M., Talmon, Y., & Zemb, T. (2021). Using microemulsions: Formulation based on knowledge of their mesostructure. *Chemical Reviews*, *121*(10), 5671–5740. DOI: 10.1021/acs.chemrev.0c00812

Hackett, J. (2014). Cleaning PVC with microemulsions. *V & A Conservation Journal*, *62*. Available at: http://www.vam.ac.uk/content/journals/conservation-journal/autumn-2014-issue-62/cleaning-pvc-with-microemulsions-joanne-hackett/

Jin, L., Budhathoki, M., Jamili, A., Li, Z., Luo, H., Shiau, B. J. B., Delshad, M., & Harwell, J. H. (2017). Predicting microemulsion phase behavior using physics based HLD-NAC equation of state for surfactant flooding. *Journal of Petroleum Science and Engineering*, *151*, 213–223. DOI: 10.1016/j.petrol.2016.12.035

Lochhead, R. Y. (2017). The use of polymers in cosmetic products. In *Cosmetic Science and Technology: Theoretical Principles and Applications* (pp. 171–221). Elsevier Inc. DOI: 10.1016/B978-0-12-802005-0.00013-6

Mejia Jr., A. B. (2015, April 16). Characteristic curvature evaluation of anionic surfactants for enhanced oil recovery applications. *Proceedings of the National Conference On Undergraduate Research (NCUR)*. National Conference On Undergraduate Research (NCUR), Eastern Washington University Cheney, WA. Available at: https://libjournals.unca.edu/ncur/wp-content/uploads/2021/07/1224-Mejia-FINAL-W-Revisions.pdf

Miller, C. (2008). Solubilization in surfactant systems. In K. Birdi (Ed.), *Handbook of Surface and Colloid Chemistry* (3rd Edition; pp. 415–438). CRC Press. DOI: 10.1201/9781420007206.ch9

Mina, L. (2020d). Foxy underpants: Or the use of chelators and enzymes to reduce foxing stains on early nineteenth century men's linen underpants. *Journal of the American Institute for Conservation*, *59*(1), 3–17. DOI: 10.1080/01971360.2019.1674604

Nakama, Y. (2017). Surfactants. In *Cosmetic Science and Technology: Theoretical Principles and Applications* (pp. 231–244). Elsevier Inc. DOI: 10.1016/B978-0-12-802005-0.00015-X

Nakamura, D., Hirano, M., & Ohta, R. (2017c). Nontoxic organic solvents identified using an: A priori approach with Hansen solubility parameters. *Chemical Communications*, *53*(29), 4096–4099. DOI: 10.1039/c7cc01434a

Ormsby, B., Keefe, M., Phenix, A., Von Aderkas, E., Learner, T., Tucker, C., & Kozak, C. (2016). Mineral spirits-based microemulsions: A novel cleaning system for painted surfaces. *Journal of the American Institute for Conservation*, 55 (1), 12–31. DOI:10.1080/01971360.2015.1120406.

Palmer, M., & Hatley, H. (2018). The role of surfactants in wastewater treatment: Impact, removal and future techniques: A critical review. *Water Research*, *147*, 60–72. DOI: 10.1016/j.watres.2018.09.039

Salager, J. L., Bourrel, M., Schechter, R. S., & Wade, W. H. (1979). Mixing rules for optimum phase-behavior formulations of surfactant/oil/water systems. *Society of Petroleum Engineers Journal*, *19*(05), 271–278. DOI: 10.2118/7584-PA

Salager, J.-L., Antón, R., Bullón, J., Forgiarini, A., & Marquez, R. (2020). How to use the normalized hydrophilic-lipophilic deviation (HLDN) concept for the formulation of equilibrated and emulsified surfactant-oil-water systems for cosmetics and pharmaceutical products. *Cosmetics*, *7*(3), 57. DOI: 10.3390/cosmetics7030057

Salager, J.-L., Marquez, N., Graciaa, A., & Lachaise, J. (2000). Partitioning of ethoxylated octylphenol surfactants in microemulsion–oil–water systems: Influence of temperature and relation between partitioning coefficient and physicochemical formulation. *Langmuir*, *16*(13), 5534–5539. DOI: 10.1021/la9905517

Sato, M., & Quye, A. (2019). Detergency evaluation of non-ionic surfactant Dehypon® LS54 for textile conservation wet cleaning. *Journal of the Institute of Conservation*, *42*(1), 3–17. DOI: 10.1080/19455224.2018.1556719

Schröer, L., Fiers, G., Deprez, M., Boon, N., Cnudde, V., Soens, L., & De Kock, T. (2022b). Examining the potential of enzyme-based detergents to remove biofouling from limestone heritage. *Coatings*, *12*(3). DOI: 10.3390/coatings12030375

Shashoua, Y., & Segel, K. (2013). Cleaning plastics in museums. *Meddelelser om Konservering*, *2*, 3–12.

Solans, C., & García-Celma, M. J. (2017). Microemulsions and nano-emulsions for cosmetic applications. In *Cosmetic Science and Technology: Theoretical Principles and Applications* (pp. 507–518). Elsevier Inc. DOI: 10.1016/B978-0-12-802005-0.00029-X

Stavroudis, C. (2012c, September). More from CAPS3: Surfactants, silicone-based solvents, and microemulsions. *WAAC Newsletter*, *34*(3), 24–27. Available at: https://cool.culturalheritage.org/waac/wn/wn34/wn34-3/wn34-3.pdf

Stavroudis, C. (2016b). *Modular Cleaning Recipes*. Available at: https://www.getty.edu/conservation/our_projects/education/caps/modular_cleaning_recipes.pdf

Stavroudis, C., & Blank, S. (1989b, May). Solvents & sensibility. *WAAC Newsletter*, *11*(1), 2–10. Available at: https://cool.culturalheritage.org/waac/wn/wn11/wn11-2/wn11-202.html

Stavroudis, C., Doherty, T., & Wolbers, R. (2005). A new approach to cleaning I: Using mixtures of concentrated stock solutions and a database to arrive at an optimal aqueous cleaning system. *WAAC Newsletter*, *27*(2), 17–28.

St-Laurent, G. (1996). *The Care and Handling of Recorded Sound Materials*. [online] Available at: https://cool.culturalheritage.org/byauth/st-laurent/care.html [Accessed 12 Jul. 2023]

Tímár-Balázsy, Á., & Eastop, D. (1998b). *Chemical Principles of Textile Conservation*. Butterworth-Heinemann.

Tiwari, S., Mall, C., & Solanki, P. P. (2020). CMC studies of CTAB, SLS & Tween 80 by spectral and conductivity methodology to explore its potential in photogalvanic cell. *Surfaces and Interfaces*, *18*, 100427. DOI: 10.1016/j.surfin.2019.100427

Tsujii, K. (2017). Wetting and surface characterization. In *Cosmetic Science and Technology: Theoretical Principles and Applications* (pp. 373–388). Elsevier. DOI: 10.1016/B978-0-12-802005-0.00023-9

Utomo, A., Alderman, N. J., Padron, G. A., & Özcan-Taşkın, N. G. (2023). A comparative study on emulsification in the presence of a nanoclay (Pickering emulsion) or a surfactant using high intensity mixing. *Chemical Engineering Research and Design*, *197*, 617–627. DOI: 10.1016/j.cherd.2023.07.041

Witthayapanyanon, A., Harwell, J. H., & Sabatini, D. A. (2008). Hydrophilic–lipophilic deviation (HLD) method for characterizing conventional and extended surfactants. *Journal of Colloid and Interface Science*, *325*(1), 259–266. DOI: 10.1016/j.jcis.2008.05.061

Wolbers, R. (2000b). *Cleaning Painted Surfaces: Aqueous Methods*. Archetype.

Yamamoto, H. (2012). *Estimation of EACN (equivalent alkane carbon number) with HSP*. Pirika Hansen Solubility Parameters(HSP) Application Notes. [online] Available at: https://pirika.com/NewHP/PirikaE3/EACN.html [Accessed 12 May. 2024]

Yamashita, Y., Miyahara, R., & Sakamoto, K. (2017). Emulsion and emulsification technology. In *Cosmetic Science and Technology: Theoretical Principles and Applications* (pp. 489–506). Elsevier Inc. DOI: 10.1016/B978-0-12-802005-0.00028-8

Yang, Y., Fang, Z., Chen, X., Zhang, W., Xie, Y., Chen, Y., Liu, Z., & Yuan, W. (2017). An overview of pickering emulsions: Solid-particle materials, classification, morphology, and applications. *Frontiers in Pharmacology*, *8*. Available at: https://www.frontiersin.org/articles/10.3389/fphar.2017.00287

Zarate-Muñoz, S., Texeira de Vasconcelos, F., Myint-Myat, K., Minchom, J., & Acosta, E. (2016). A simplified methodology to measure the characteristic curvature (Cc) of alkyl ethoxylate nonionic surfactants. *Journal of Surfactants and Detergents*, *19*(2), 249–263. DOI: 10.1007/s11743-016-1787-x

Zemb, T. N., Klossek, M., Lopian, T., Marcus, J., Schöettl, S., Horinek, D., Prevost, S. F., Touraud, D., Diat, O., Marčelja, S., & Kunz, W. (2016). How to explain microemulsions formed by solvent mixtures without conventional surfactants. *Proceedings of the National Academy of Sciences*, *113*(16), 4260–4265. DOI: 10.1073/pnas.1515708113

# Index

Pages in **bold** refer to tables.

FSC
www.fsc.org
MIX
Papier | Fördert
gute Waldnutzung
FSC® C083411

Kolibri 360 GmbH
Ferdinand-Jühlke-Straße 7
99095 Erfurt, Deutschland
produktsicherheit@kolibri360.de